FIRE CAPTAIN ORAL EXAM STUDY GUIDE

2nd edition

1995
BY
ARTHUR
R.
COUVILLON

Printed in the United States of America

Library of Congress Cataloging Data:

Couvillon, Arthur R.
"Fire Captain Oral Exam Study Guide"
Fire Service Entrance Exam Preparation, text books, etc.
Fire Service: Handbooks, manuals, text books, etc.
Fire Engineer: Handbooks, manuals, text books, etc.
Fire Captain: Handbooks, manuals, text books, etc.
Firefighter: Handbooks, manuals, text books, etc.
Information: Handbooks, manuals, text books, etc.
Promotional: Handbooks, manuals, text books, etc.
Practice Exams: Fire Service + Paramedics + EMT
EMT & Paramedic Recertification Study Guides
I. Title
ISBN 0-938329-57-X
LCCN 94-079291

INTRODUCTION

This "STUDY GUIDE" is a compilation of typical "ORAL INTERVIEW QUESTIONS", "ASSESSMENT CENTERS", and "INCIDENT SIMULATOR" with emphasis directed towards, Fire Captain, and/or Fire Lieutenant.

Most Fire Departments have an "ORAL INTERVIEW" as part of their promotional process. Many Fire Departments conduct the "ORAL INTERVIEW" as a portion of an "ASSESSMENT CENTER". Also included in many promotional exams for the position of Fire Captain, and/or Lieutenant are "INCIDENT SIMULATOR TEST". All three types of test are covered in this "STUDY GUIDE".

This "STUDY GUIDE" is designed to prepare Fire Captain, and/or Fire Lieutenant candidates for the "ORAL INTERVIEW", "ASSESSMENT CENTER", and "INCIDENT SIMULATOR" portions of the promotional process.

The use of this "STUDY GUIDE" can reduce the element of surprise in the "ORAL INTERVIEW", "ASSESSMENT CENTER", and during the "INCIDENT SIMULATOR TEST", by exposing prospective candidates to possible questions and responses for the "ORAL INTERVIEW" and also exposing the prospective candidates to typical formats for "ASSESSMENT CENTER" and examples of incident situations used in "INCIDENT SIMULATOR TEST".

The principal intent of this "STUDY GUIDE" is to motivate each candidate to consider the possible questions and/or situations that may arise in any portion of the promotional process.

ACKNOWLEDGEMENTS

Appreciation is expressed to those individuals, Firefighters, Fire Engineers, Fire Captains, Fire Training Officers, Fire Chiefs, Fire Technology Instructors, and Personnel Directors that have contributed to the compilation of this book: **"FIRE CAPTAIN ORAL EXAM STUDY GUIDE"**. A special thanks to Chief Ralph Mailloux of the GFD for the design and drawing of the cover for this book!

ABOUT THE AUTHOR

Arthur R. Couvillon is a veteran Firefighter with over 25 years of experience. Art shares the knowledge and study techniques that he has gained during this period along with research and surveys that he has conducted with Fire Service authorities all across the U.S.A. Listed below are some of Art's published "Fire Service Career" books:

THE COMPLETE FIREFIGHTER CANDIDATE
FIREFIGHTER WRITTEN EXAM STUDY GUIDE
FIREFIGHTER ORAL EXAM STUDY GUIDE
FIRE SERVICE ENTRANCE EXAM PREPARATION
ADVANCEMENT IN THE FIRE SERVICE
FIRE ENGINEER WRITTEN EXAM STUDY GUIDE
FIRE ENGINEER ORAL EXAM STUDY GUIDE
FIRE SERVICE HYDRAULICS
FIRE CAPTAIN WRITTEN EXAM STUDY GUIDE
FIRE CAPTAIN ORAL EXAM STUDY GUIDE
FIREFIGHTER WRITTEN PRACTICE EXAMS
FIRE ENGINEER WRITTEN PRACTICE EXAMS
FIRE CAPTAIN WRITTEN PRACTICE EXAMS
EMT RECERTIFICATION STUDY GUIDE
EMT RECERTIFICATION PRACTICE EXAMS
PARAMEDIC RECERTIFICATION STUDY GUIDE
PARAMEDIC RECERTIFICATION PRACTICE EXAMS

TABLE OF CONTENTS

SECTION 1

THE ORAL INTERVIEW

WHY THEY HAVE ORAL EXAMS

The purpose of an **ORAL INTERVIEW** is to evaluate each candidates personal qualification, training, experience, attitude, personal attributes, and any intangibles involved that usually outline a candidates likelihood of success or failure for the position of **FIRE CAPTAIN**.

The objective of the **ORAL INTERVIEW** is to determine and identify various elements that have not been examined in the other phases of the exam process. Some examples are:

1. Adaptability, Stability, Poise.
2. Attitude, Enthusiasm, Integrity.
3. Ability to express yourself.
4. Ability to function under stressful situations.
5. Ability to work with others.
6. Ability to follow directions.
7. Your judgement.
8. Your attitude toward supervision.
9. Your personal grooming habits. ETC.

Remember that you asked for this opportunity to be interviewed when you filled for your application. You are not here against your will, and can choose to withdraw at any time.

The decision to be or not to be interviewed is yours to make. You are being interviewed because thus far you have shown the basic qualifications, technical and/or intellectual abilities, along with the experience being sought. You are still in the running and this is an important step in the selection process.

HAVE A POSITIVE ATTITUDE CONCERNING THE ORAL INTERVIEW, BE ENCOURAGED THAT YOU ARE STILL IN THE RUNNING!

TYPES OF ORAL BOARDS

Fire Captain-Lieutenant oral boards usually consist of three members that may be selected from:

1. The community.
2. Business and industry.
3. Fire Department members.
4. Members from other Fire Departments.
5. Minority group representatives.
6. Personnel Department.

Oral interview boards are set up in different areas in different ways and are composed of various groups of interviewers. In some cases the oral boards are composed entirely of lay members of the Department for which the test is given. Some interview boards are composed of members from other Fire Departments.

There is no set standard or criteria as to where, who, how many, or from what positions the board may be chosen from, although certain standards are attempted.

ORAL INTERVIEWS may be one of two types:
1. Stressful. 2. Non-Stressful.

STRESSFUL ORAL BOARDS will attempt to evaluate you by creating a stressful atmosphere so as to see how you will perform under this stress. In this interview without introduction you will be asked a question which requires taking a stand on a controversial subject, and then all of the board members will take the opposite view point. The chairman of the board or one of the members previously designated by the board, will inject new controversies on different subjects at random intervals, never letting you fully explain your position on any subject. These types of interviews are seldom used.

NON-STRESSFUL ORAL BOARDS (ordinary interview) are normally of a cordial atmosphere where the board will attempt to put you at ease during the complete interview. This interview usually starts with some type of introductory remark that is designed to reassure you, and is followed by some review questions that you are familiar with. These types of questions are designed to show your interest, vocabulary, along with the volume and tone of your voice, your mannerisms, etc.

HOW TO PREPARE FOR THE ORAL INTERVIEW

You have been preparing for the **ORAL INTERVIEW** portion of the exam from the time that you started to learn about the Fire Service as a career.

Some of the **PRE-ORAL INTERVIEW** preparation that you have already established include:

1. Submission of your application and resume will have informed the oral board as to your sincerity and desire to obtain the position, along with showing your ability to organize your thoughts, your neatness, and your background. This is the first impression that the oral board will have of you, make it good!
2. Knowledge gained while acting in the position.
3. Knowledge gained concerning the position.
4. Knowledge gained concerning the Department.
5. Knowledge gained concerning the community.
6. Education and training.
7. Knowing your competition.
8. Knowing who is getting promoted.
9. Self assessment.
10. Your exam checklist.
11. Practice exams/interviews.

ORAL INTERVIEW EXAM CHECKLIST

1. Know what is on your application.
2. Know what is on your resume.
3. Know the duties of the position.
4. Know the responsibilities of the position.
5. Know the qualifications of the position.
6. Know the Department.
7. Be prepared to present yourself effectively.
8. Know what you have done.
9. Know what you can do.
10. know what the job requires in the way of performance.
11. Visualize questions that may be asked.
12. Visualize answers to questions.
13. **PRACTICE**.

THE ORAL INTERVIEW

The **ORAL INTERVIEW** for Fire Captain-Lieutenant occasionally will last less than 15 minutes, the average time is from 20 to 30 minutes, but may last longer!

On the day of the **ORAL INTERVIEW**:

1. Arrive early, know where you are going, where you are to park in advance, allow time to get there, park, and reach the interview room, announce your arrival to the proper person, find a place to sit and relax until you are called.
2. Bring your application for review, or some light reading.
3. Do not bring any exhibits or technical material unless you have been instructed to do so.
4. Be clean and well groomed.
5. Be neat.
6. Avoid all extremes in dress and hair style.
7. Do not wear any pins or emblems.
8. Suit, or sport coat and tie are recommended.
9. While waiting to be called, recheck your attire
10. While waiting restudy your application.

When you are called for the interview, remember that the interview is a sales interview and that you are the salesman and that the board is the client, and that the product that you are selling is you.

Upon entering the interview room, politely acknowledge any introductions which may be made before you sit down.

If you know a member of the board do not try to hide it, but you do not have to emphasize it either.

Usually the interview board will be seated on one side of a table and you will see an empty chair for you on the opposite side of the table, but there is no set policy regarding the seating arrangements.

Enter the room standing erect and walk with confidence.

You will be introduced to each member of the board, as you are introduced, look each in the eyes and acknowledge along with a firm hand shake if the opportunity is extended. Be confident and direct, use rank titles if applicable.

If the board members are introduced by name, try to remember the names and where possible address the members by name during the interview and/or at its conclusion. Sirs will be satisfactory if you think that you may become confused. Do not confuse title or rank between board members.

You will now be asked to take a seat, sit erect in the chair, do not seem stiff and rigid, be erect and comfortable. Don't shift positions constantly.

It is alright to gesture with your hands, but not too much! You do not want the board staring at your hands. Do not sit on your hands, put your hands in your lap or on the table in front of you. Be natural in your expression and movements. Avoid distracting movements, such as scratching, pulling on buttons etc.

When asked questions look at the face of the person asking the question as this will help you to focus on the question. When answering a question, direct the answer to the person that asked the question, but do not ignore the other members of the board.

The board will usually review your application at the start of the interview. Do not interrupt unless there is a significant error made, do not quibble over matters of minor importance.

After the board reviews your application, the first question usually will be a question such as: Tell us something about yourself or Why do you want the position of Fire Captain-Lieutenant? Try to have some opening statement prepared about yourself, your background and qualification and why you want to be a Fire Captain-Lieutenant. Be careful here, never memorize and recite a prepared statement, just have an organized list in your head that you would feel comfortable elaborating on.

Do not give one word answers to questions, try to let the board know why an answer is "yes" or "no", sell yourself!

Answer questions confidently and to the best of your ability. Do not try to "fake it", be honest, if you do not know an answer to a question tell the board that you do not know the answer. If you lie, you can be disqualified! Be sincere and deliberate with your answers.

Remember that you want this position, sell yourself to the board, don't sell yourself short. The board members are trying to select the best man for the position, you must give them something to work with.

Don't worry if you are nervous, the board expects this. Most board members usually will want you to do your best. The board members are not your enemies, they are just observers looking for the best candidate for the position. Let the board know that you are the best candidate for the position. Also let the board know that you are a serious candidate, act professional at all times during the interview. If appropriate a little humor is alright, but do not overdo it. Be attentive.

WHEN ANSWERING QUESTIONS:

1. Modulate your voice.
2. Speak-up, but not too loudly.
3. Make sure that you understand the question before answering, if not, restate the question or ask for a clarification. Don't overdue this, do not make the board repeat every question.
4. Be honest.
5. Be pleasant, smile occasionally.
6. Wait until the entire question is asked, don't interrupt the questions.
7. Do not argue. If you make a stand on an issue, do not change your opinion unless you are proven wrong.
8. Answer questions completely and then stop, don't try to dominate the interview.
9. Expect abrupt changes in the questioning.
10. Don't wisecrack.
11. Reply with your answers promptly, not hastily.
12. Don't spend too much time praising your position.

Sometimes a board member may stop you in the middle of answering a question, this could be a positive indication that he feels that you have already answered the question to his liking and he wants to cover another area. Never try to continue to answer the question! Be ready for the next question.

Just prior to the end of the interview, the board will usually ask if there is anything that you would like to add:

1. If they have overlooked one or more of your strong points, you should bring it out now.
2. Do not start an extended presentation at this point.
3. If you have nothing to add, just say; No thank you, I believe we have covered everything.
4. Don't compliment the board members.

The board will let you know when the interview is over, at this time just thank them, shake their hands and leave the room. Do not make a speech at this time. The only time that you should say anything is if something very important has been left out, then you should briefly mention it. You actually terminate the interview, this is the point where you can actually talk yourself out of a good impression or fail to present an important bit of information.

When you leave the interview room you should leave with poise and confidence watch for an offer of a handshake, if an offer is made, shake hands.

HOW TO PUT YOUR BEST FOOT FORWARD

Throughout this process, you may feel that the oral board is trying to penetrate your defenses,to seek out your hidden weaknesses, and to try to embarrass or confuse you. This is not the case, the are compelled to make an evaluation of your qualifications for the position. They want to see you at your best.

The oral board must interview each candidate and a noncooperative candidate may become a failure in spite of the boards best efforts to bring out the candidates qualifications. You must put your best foot forward, the following are some suggestions on how to accomplish this task:

BE NATURAL: keep your attitude confident, but not cocky. If you are not confident that you can do the job, don't expect the board members to be. Don't apologize for your weaknesses, bring your strong points. The board is interested in a positive, not a negative presentation. Cockiness will antagonize any board member, and make him wonder if you are covering up a weakness by a false show of strength.

GET COMFORTABLE: don't lounge or sprawl, sit erectly but not stiffly. A careless posture may lead the board to conclude you are careless in other things, or at least that you are not impressed by the importance of the interview. Don't fuss with your clothing or with a pencil or ash tray, etc. Your hands may occasionally be useful to emphasize a point; don't let them detract from your presentation by becoming a point of distraction.

DON'T WISECRACK: or make small talk, this is a serious situation, and your attitude should show that you consider it as such. Further, the time of the board is limited; they don't want to waste it, and neither should you.

IF YOU KNOW A MEMBER OF THE BOARD: don't make a point of it, but don't hide it. Certainly you are not fooling him, or the other members of the board. Don't try to take advantage of your acquaintance with this individual, it will likely hinder your score.

DON'T ATTEMPT TO DOMINATE THE INTERVIEW: Let the board have control. They will give you clues, don't assume that you have to do all of the talking.

DON'T EXAGGERATE: your experience or abilities, the board has your application in front of them and also from other sources may know more about you than you think. Also you probably won't get away with it anyway. An experienced board is rather adept at spotting such a situation. Don't take a chance!

DON'T BE TECHNICAL OR BORING: The board is not interested in ponderous technical data at this time.

BE AWARE that the board has a number of questions to ask you. Don't try to take up all the interview time by showing off your extensive knowledge of the answer to the first question that you are asked.

BE ATTENTIVE: you will have a limited time in the interview, keep your attention at a sharp level throughout this period. When a board member is questioning you, give this person your undivided attention. Direct your response primarily to the board member that ask the question, but don't exclude the other board members.

DON'T INTERRUPT: A board member may be stating a problem for you to analyze. When the time comes you will be asked a question. Let the board member state the problem, and wait for the question.

MAKE SURE THAT YOU UNDERSTAND THE QUESTION: Don't try to answer until you are sure what the question is. If the question is not clear, restate it in your own words or ask the board member to clarify it for you. Don't haggle about minor elements.

REPLY PROMPTLY: But don't reply hastily. A common entry on oral board rating sheets is " candidate responded readily" or "candidate hesitated in replies". Respond as promptly and quickly as you can, but don't jump to hasty responses.

DON'T BE PREEMPTORY IN YOUR ANSWERS: A brief answer is proper, but don't fire your answers back. This is a losing game from your point of view. The board member can probably ask questions much faster than you can answer them.

DON'T TRY TO CREATE ANSWERS THAT YOU ASSUME THAT THE BOARD WANTS TO HEAR: The board is interested in what kind of a mind you have and how it works, not in playing games. Most board members can spot this tactic and will grade you down for it.

DON'T BE INGRATIATING: This routine rarely works with an oral board. Be pleasant and smile occasionally, but do it naturally and don't overdo it.

DON'T CHANGE VIEWS IN ORDER TO PLEASE THE BOARD: Board members will take a contrary position in order to draw you out and see if you are willing and able to defend your point of view. Don't start a debate, don't surrender a good position. If your position is worth taking, it is worth defending.

IF YOU ARE SHOWN TO HAVE MADE AN ERROR IN JUDGEMENT, DON'T BE AFRAID TO ADMIT THE ERROR: The board that you are forced to reply without any opportunity for careful consideration. Your answer may be demonstrably wrong. If so, admit it and get on with the interview.

DON'T SPEND TOO MUCH TIME DISCUSSING YOUR PRESENT JOB: The opening statements may concern your present employment. Answer the question but don't go into an extended discussion. You are being examined for a new job, not your present one. As a mater of fact, try to phrase all you answers in terms relating to the position of Firefighter.

DON'T TELL STORIES: Keep your responses to the point. If you feel the need for illustration from your personal experience, keep it short. Leave out the minor details. Make sure that the incident is true.

DON'T USE SLANG TERMS: many a good response has been weakened by the injection of slang terms or other jargon. Oral boards usually will notice any slips of the grammar or any other evidence of carelessness in speech habits.

DON'T BRING DISPLAYS OR DEMONSTRATIONS: The board members are not interested in letters of reference etc.

SOME WAYS THAT CAN CAUSE YOU TO "STRIKE OUT" IN AN ORAL INTERVIEW:

1. Poor personal appearance.
2. Lack of interest and enthusiasm.
3. Passiveness or indifference.
4. Overemphasis on wages.
5. Condemnation of past employers.
6. Failure to look at board members during interview.
7. Limp, fishy handshake.
8. Indefinite response to questions.
9. Overbearing, overaggressive, conceited with superiority or "know it all" attitude.
10. Inability to express self clearly: poor voice, diction, or gammer.
11. Lack of planning for position.
12. Lack of confidence and poise: nervous, ill at ease.
13. Make excuses: evasive; hedges on unfavorable factors in work record, etc.
14. Lack of tact.
15. Lack of courtesy; ill mannered.
16. Lack of maturity.
17. Lack of vitality.
18. Indecision.
19. Sloppy application.
20. Merely "shopping" for position.
21. Wants position only for a short time.
22. Lack of interest in jurisdiction.
23. Appear lazy.
24. Low moral standards.
25. Intolerant: strong prejudices.
26. Narrow interest.
27. Inability to take criticism.
28. High pressure type.
29. Inability to listen.
30. Domination of the interview.

20 SUGGESTIONS FOR PUTTING YOUR BEST FOOT FORWARD IN THE ORAL INTERVIEW:

1. Be natural.
2. Get comfortable, but don't lounge or sprawl.
3. Don't "wisecrack" or make small talk.
4. Don't exaggerate your experience or abilities.
5. If you know a member of the board, don't make a point of it, but don't hide it.
6. Don't dominate the interview.
7. Be attentive.
8. Don't interrupt.
9. Make sure you understand the question.
10. Reply promptly but not hastily.
11. Don't be preemptory with your answers.
12. Don't try to create the answer that you think that the board members want to hear.
13. Don't switch sides in your reply merely to agree with a board member.
14. Don't be afraid to admit an error in judgement if you are shown to be wrong.
15. Don't dwell at length on your present job.
16. Don't bring in extraneous comments or tell lengthy anecdotes.
17. Don't be too technical or ponderous.
18. Don't use slang terms.
19. Leave your exhibits at home.
20. Don't be flattering.

ADDITIONAL ORAL INTERVIEW INFORMATION

Oral interviews of candidates for Fire Service promotional exams has been accepted as part of the over all test procedure that determines the candidates passing and the candidates position placement on the eligibility list.

Much has been said pro and con relative to the merits of the oral interview, but the candidates must accept it, realizing that whereas it is not perfect, it is certainly a step forward in the merit system.

Oral interview boards are set up in different areas in different ways and are composed of various groups of interviewers. In some cases the oral boards are composed entirely of lay members of the department for which the test is given. Some interview boards are composed of members of other departments of other cities. There is no set gauge as to where, who, how many or from what positions the board may be chosen from, although certain standards are attempted.

The oral interview may last from fifteen minutes to forty-five minutes or longer. The average is about twenty-five to thirty-five minutes.

The candidate who is preparing for an examination cannot afford to overlook preparation for the oral interview. Preparation should begin long before the candidate has taken the written portion of the promotional exam. Often the oral exam is given within a day after the written exam. The proper preparation and training for an oral test requires considerable time. Of course, some candidates are better than others and will not require the same amount of time to prepare. Don't neglect to prepare for the oral prior to the written.Don't wait until you have taken the written to determine if you will pass or not!

Whatever your general and special qualifications may be, or however well you may pass the written examination, the hurdle of the oral interview is still ahead. Your acceptance or rejection will depend largely upon it. Your success in the interview will depend largely upon how you handle yourself.

When the interview time is approaching, if you are normal you will feel a certain amount of uneasiness about the interview. How will you be received? What will you be asked? How can you possibly prepare for the interview when you don't even know the line of attack that the interviewers will use? You may feel, as many do, that this is a highly unequal contest; all the weapons, strategy, and imitative are stacked in favor of the interviewers. What if you fail? Maybe you would be wise to withdraw your application and wait for another opportunity.

LOOK AT YOUR ATTITUDE OF THE INTERVIEW:

First, what is the reason for an interview? Is it to test your technical knowledge? In some types of interviews such as Fire Engineer, yes many technical questions will be asked. The main purpose of the interview is to bring out certain intangible qualities which cannot be easily measured otherwise, and to place the applicant in a listing of many candidates, each ultimately to be weighed against the others.

Second, why are you to be interviewed? You asked for this opportunity when you filed for your application. You are not here against your will, and can choose to withdraw even now.
The decision to be, or not to be interviewed is yours to make. You are being interviewed because thus far you have shown the basic qualifications, the technical and/or intellectual abilities, and the experience being sought. You are still in the running, and this is an important step in the selection process.

You're uneasy now mostly because you do not know what to expect, and don't know what you can do in advance of, or during the interview to make the best of it. But remember, usually the interviewer or interviewing board has sufficient competence and experience in interviewing, to put you at ease and to permit you to demonstrate those intangible qualities for which they are looking. Still, there are several specific things you can do in advance to improve your performance in the ORAL INTERVIEW situation:

1. Keep a copy of your application and review it carefully before the interview. Know what you have written, and if asked about any point covered in the application, such as your work experience, you will have no trouble or uncertainty about what you said and will not fumble for an answer.

2. Study the specifications and qualifications set forth in the job announcement. An applicant for promotion must thoroughly know the duties and responsibilities of the job he seeks. You must also be prepared to solve hypothetical problems, or answer questions that will reveal your specific understanding of principles and philosophy of the job.

3. Think through each qualification required. How would you go about questioning someone to assure yourself that they are qualified? How would you answer those questions? Don't kid yourself! Be frank and realistic in seeking out the holes in your armor.

4. Do some general reading in areas in which you feel that you may be weak. Don't try to memorize statistics, procedures, and/or definitions at this late date. The main advantage of general reading is that it may help you to consolidate and organize your own thinking and experience.

5. Watch your health, and particularly your mental attitude. Get plenty of sleep if you possibly can.

When the interview day is here, plan to arrive early, allowing plenty of time to get there, park your car, reach the interview room, and find a place to sit and relax before you are called. Take a long some light reading, or your application form to review, but don't go loaded with exhibits or technical materials.

Dress your best, but not flashily. Be well groomed; scrubbed, pressed, trimmed, and as immaculately fresh as possible. Do not wear pins or identification marks. Soon your name will be called. From here you are on your own.

Acknowledge any introduction before you sit down. The interviewer or chairman will probably make some introductory remark, and may begin the interview with review questions; your education, work experience, training, and/or other qualifications. As questioning continues, you can expect all board members present to participate.

Don't start off on a extended extemporaneous speech just because you may be asked if you have anything to add. This may be asked to give you the opportunity to mention some significant fact that may not have been touched upon. Cover it as briefly as you can. If they want to know more about a point they will ask you.

During the questioning you may feel that the interviewer, or some member of the board, is trying to seek out your hidden weakness, to confuse you, or keep you on the defense. Remember that this isn't so! By pushing points in some detail the interviewer learns something about your understanding of principles, your ability to think and express yourself clearly, your judgement, your convictions, and something about your temperament, attitudes and poise. However irritating some questions may be, make sure your answers are sincere and purposeful.

ORAL INTERVIEW SUGGESTIONS TO KEEP YOU ON COURSE:

1. GET COMFORTABLE: if you know any member present, don't advertise it, but don't hide it either. When acknowledging introductions, address the interviewers by their appropriate title. Sit comfortably, erect and relaxed, rather than stiffly braced. If you are holding something, keep your hands loosely clasped. Don't fumble with trinkets or personal items.

2. BE YOURSELF: be pleasant but not chummy. Avoid "wisecracking" and artificial grinning and/or pleasantness. Watch your grammar and avoid slang expressions, but good English doesn't require exaggerated elocution or stilted language. Be confident, but not cocky. Don't exaggerate your experience or training. The board already knows quite a bit about you, and almost all interviewers are experts at spotting phonies.

3. BE ATTENTIVE: time is limited, so don't waste it. Don't give all your attention to one member as all are going to pass on you. Different members may be using different approaches in questioning, so be alert for a change of tactics or change of subject. If a member is starting a lengthy problem for you to analyze, do not interrupt or "jump the gun", let him state his question fully before replying.

4. THINK BEFORE ANSWERING: reply promptly but not hastily. If you do not understand the question, rephrase it or ask the member to explain it. Inappropriate or unorganized replies usually come from talking before you have considered your reply carefully. here a qualified answer is needed, give as such, but do not straddle the fence where a positive stand is indicated. If the interviewer challenges you on a point, don't surrender a good position in a scramble for cover; he may be testing your convictions or your logic. Don't be dogmatic, if your wrong, admit it!

5. KEEP THE PURPOSE OF THE INTERVIEW IN MIND: leave exhibits, clippings, articles you have prepared, etc home. Unless you have asked to been bring them don't bring them! void trying to match wits with the interviewers, and especially don't try to second guess him or create the answer you think is wanted. Be objective; don't be/seem antagonized or show your resentment if a member becomes critical, as he is probably doing it for a purpose.

6. BE BRIEF: don't be technical or ponderous, and especially avoid "professional jargon". If personal experience is needed for illustration, come directly to the point, and stick to real incidents. Don't dominate the conversation with long historical explanations or by trying to show all you know about a subject. Let the interviewer lead you; if he wants to know more, he'll ask you. When the interview is ended, take your leave courteously but promptly; give a glance around the room and a courteous "thank you, gentlemen" is all that you need to say.

Perhaps you will forget many of the "do's" and "don'ts" when your interview starts. Even remembering them all will not assure you a high rating. The interviewer, or board, wants you to do your best, but they have a heavy responsibility. How you respond under pressure, the poise you maintain, the personal traits you show, and your general manner of responses are testing devices, only if you are under pressure. They do not want an idealized portrait of you, but a representative photograph of you as you really are, plus a X-ray of things that may not be apparent on the outside. It's up to you to show the interviewers or board that you really possess the qualifications that they are looking for. Proper preparation may help you to do this.

8 POINTS CANDIDATES SHOULD DO PRIOR TO THE ORAL INTERVIEW:

1. Keep a copy of your application and review it carefully before the interview.
2. Check as to the members of the board and with other candidates as to the type of questions asked.
3. Study the class specifications and the total examination announcement.
4. Think through each qualification required.
5. Do some general reading in areas in which you feel that you may be weak.
6. Drill yourself with possible questions/answers.
7. Get a good nights sleep and watch your general health and mental attitude.
8. Dress neat and clean, not flashily. No wearing of identification pins.

TYPICAL ORAL INTERVIEW QUESTIONS

Some examples of Fire Captain-Lieutenant oral exam questions that you should be familiar with are:

1. Tell us something about yourself.
2. How have you prepared yourself for the position of Fire Captain -Lieutenant?
3. Why do you want the position of Fire Captain-Lieutenant?
4. What are your qualifications for the position of Fire Captain-Lieutenant?
5. Why should you be the first candidate chosen?
6. What are your immediate and ultimate goals in the Fire Service?
7. When have you decide to promote to Fire Captain-Lieutenant?
8. What is your most outstanding asset?
9. What is your biggest weakness?
10. Is there anything that you would like to add?

SECTION 2

PERSONAL INFORMATION

QUESTION:

Tell us something about **YOURSELF**.

POSSIBLE RESPONSE:

I am ___ years old, married/unmarried with ___ children. I have been a Firefighter with the ___________ Fire Department for ___ years, of which ___ years I was a Paramedic, and ___ years as a Fire Inspector, currently I am a Fire Engineer. During these years, through **MOTIVATION** and **INVOLVEMENT** in the Department, I have gained **EXPERIENCE** and the **KNOWLEDGE** necessary to prepare myself for the position of Fire Captain and/or Fire Lieutenant.

EXAMPLES:

1. ____ years as a Firefighter.
   A. Firefighting techniques.
   B. Company fire prevention procedures.
   C. Public relations.
   D. Rules and regulations.
   E. Station maintenance.
   F. Reports and records.
   G. Hydrant maintenance.
2. ____ years as a Paramedic.
   A. Normal and emergency driving situations.
   B. Responsibility of emergency situations.
   C. Decision making.
   D. Responsibility of apparatus and equipment.
   E. Rescue techniques.
   F. Records keeping and reports. (paper work)
   G. Delegation of task.
3. ____ years in Fire Prevention Bureau.
   A. Responsibility of fire prevention areas.
   B. Familiarization of codes and ordinances.
   C. Public relations.
   D. Planning and Training
   E. Reports and records.
   F. Arson.
4. ____ years as a Fire Engineer.
   A. Apparatus driving and equipment maintenance.
   B. Responsibility of apparatus and equipment.
   C. Familiarization of apparatus and equipment.
   D. Familiarization of water supply.
5. Currently in the position of Fire Engineer.
6. Currently certified as an acting Fire Captain.
7. Currently on Fire Captains Promotional list.

ADD OR DELETE APPROPRIATE EXAMPLES

QUESTION:

In order that we may all feel more comfortable, even though we have your application/resume in front of us, and have reviewed it, please give us a **BRIEF** history of your background on the __________ Fire Department.

POSSIBLE RESPONSE:

SEE RESPONSE TO PREVIOUS QUESTION!

QUESTION:

What do **YOU** consider to be your greatest **STRENGTH**?

POSSIBLE RESPONSE:

My greatest **STRENGTH** comes from my experience and ability to supervise a fire company and the personal involved, including:

1. Firefighting ability.
2. Firefighting knowledge.
3. Ability to teach and work with people.

ADD SOME OF YOUR OWN IDEAS

QUESTION:

How would **YOU** describe yourself?
How would your **PEERS** describe you?

POSSIBLE RESPONSE:

I would describe myself as an individual that shows an immense interest and respect for people and the Fire Service. It is very easy for me to get along with other people.

I believe that my **PEERS** think of me as a person that gets along with people, and as a person that has a great deal of job knowledge and interest.

ADD WHATEVER MAY APPLY TO YOU

QUESTION:

Why did you choose the fire service for **YOUR CAREER**?

POSSIBLE RESPONSE:

I chose the Fire Service as a **CAREER** for several reasons, including:

1. Security, adequate pay and benefits.
2. An opportunity for advancement.
3. For new and different challenges.
4. The variety of interesting and exciting work while at the same time working with and helping other people.
5. Pride in what I am achieving.

I believe that the Fire Service offers all of my personal requirements for a career.

ADD AND/OR DELETE WHAT IS APPROPRIATE FOR YOU

QUESTION:

In what kind of **WORK ENVIRONMENT** are you the most comfortable?

POSSIBLE RESPONSE:

I am comfortable in many different kinds of environments. I am the most comfortable in an environment where I have the opportunity to **INTERACT WITH PEOPLE**.

RESPOND WITH WHAT IS APPROPRIATE FOR YOU

QUESTION:

How do you work under **PRESSURE**?

POSSIBLE RESPONSE:

I have a personal policy to work under pressure the same way that I **NORMALLY** do under normal conditions. That is to think clearly and react systematically. In the Fire Service it is important to remain calm during many varied type of circumstances.

RESPOND WITH AN APPROPRIATE ANSWER FOR YOU

QUESTION:

Describe the most significant **CONTRIBUTION** you have made to the fire service, and how has this prepared you to be a Fire Captain?

POSSIBLE RESPONSE:

At this stage of my career my most significant contribution is in the area of **TRAINING**. I have compiled several study manuals for various apparatus and equipment. Also I have published four books dealing with promotional exams within the Fire Service.
RESPOND WITH WHATEVER IS APPLICABLE FOR YOU

QUESTION:

How have you **PREPARED** yourself for the position of Fire Captain and/or Fire Lieutenant?

Describe motivation, education, training and experience you have had to qualify you to perform the duties of a Fire Captain and/or Fire Lieutenant!

POSSIBLE RESPONSE:

I have prepared myself for the position of Fire Captain through **MOTIVATION** and **INVOLVEMENT** in the __________ Fire Department. I have also prepared myself through **TRAINING** and **EDUCATION** along with reading and serious study. During my ___ years as a Firefighter, ___ years as a Paramedic, ___ years as Fire Inspector, and ___ years as a Fire Engineer, I have gained the necessary **EXPERIENCE** and **KNOWLEDGE** for the position of Fire Captain.
GIVE EXAMPLES OF PARTICULARS THAT APPLY TO YOU!

QUESTION:

What motivates you to put forth your **GREATEST EFFORT**?

POSSIBLE RESPONSE:

I am motivated to put forth my greatest effort by the **SATISFACTION** attained from the completion of task.

QUESTION:

Do you honestly feel that you have the **CREDENTIALS** to succeed as a Fire Captain? Why?

POSSIBLE RESPONSE:

YES, I have spent a great deal of time in preparing myself for the position of Fire Captain. My qualifications extend from five distinct sources:

1. Motivation.
2. Training.
3. Education.
4. Experience.
5. Job knowledge.

MOTIVATION:

I have a strong desire to improve myself and the ________ Fire Department through involvement in many various department task and projects.

TRAINING:

My training includes reading and very serious study along with inside and outside department drills and classes, including State Officer Certification.

EDUCATION:

My education includes a College degree in Fire Science, including 28 units in Fire Science, along with completion of various Fire Service courses. I am currently enrolled in two Fire Science courses.

EXPERIENCE:

Give the number of years that you have served in the Fire Service, along with the type of experience you have gained during this period of time. EXAMPLES:

1. Firefighter.
2. Fire Paramedic.
3. Fire Inspector.
4. Arson Investigator.
5. Fire Engineer.
6. Acting Fire Captain.
7. Worked in dispatch.
8. Worked in high rise areas.
9. Worked in high incident areas.
10. Worked in the harbor area.

11. Have been responsible for:
    Apparatus and equipment.
    Subordinates.
    Emergency situations.
    Small and large fires.
    Haz Mat incidents.
    High Rise incidents.
    All phases of Fire Prevention.
    Training of subordinates.
    Quarterly task.
    Delegation of task.
    Assignment of personnel and task.
    Morale.
    Management.
    Demonstrations.
    Supervision.
    Rules and regulations.
    Policy and procedures.
    Arson investigation.
    Budgeting
    Public relations.

ADD OR DELETE WHATEVER IS APPROPRIATE FOR YOU

QUESTION:

Why do **YOU** want the position of Fire Captain?

POSSIBLE RESPONSE:

I want the position of Fire Captain for the **NEW** and **DIFFERENT CHALLENGES**, along with the **INCREASED RESPONSIBILITIES** that a Fire Captain assumes in relation to firefighting, subordinates, and other task. The position of Fire Captain will offer me the **OPPORTUNITY** to **IMPROVE** myself and the department, by allowing me to **APPLY** my **KNOWLEDGE** and **EXPERIENCE**.

ADD ANYTHING THAT MAY BE APPLICABLE FOR YOU!

QUESTION:

How have you **CONTRIBUTED** to your Fire Department?

POSSIBLE RESPONSE:

During the ___ years that I have been a Firefighter with the __________ Fire Department I have contributed in **MANY WAYS**.

EXAMPLES:

1. Involved in designing and fabricating ways to carry various tools and equipment on various apparatus. GIVE EXAMPLES!
2. Involved, designing and writing specifications for fire and rescue apparatus. GIVE EXAMPLES!
3. Researched and budgeted for various items of tools and equipment. GIVE EXAMPLES!
4. Prepared lesson plans for various tools and equipment. GIVE EXAMPLES!
5. Involved in safety projects. GIVE EXAMPLES!
6. Involved in various maintenance and overhaul of various apparatus and equipment.EXAMPLES!
7. Prepared training manual for departments Fire Engineer and Fire Captain certification programs.
8. Prepared a training manual for the proper operating procedures of our departments Ladder Trucks.
9. Prepared a training manual for the proper procedure for departments Fire Prevention program.
10. Projected a positive image of the department, while giving lectures and demonstrations of various equipment to various groups within the city.
11. Participated in the departments Paramedic program as a paramedic, and as an instructor for CPR and EMT training.
12. Participated in the departments school training program.

ADD OR DELETE ANYTHING THAT MAY OR MAY NOT APPLY!

QUESTION:

Explain any contributions or special projects that you have undertaken for the betterment of the __________ Fire Department.

POSSIBLE RESPONSE:

SEE PREVIOUS RESPONSE

QUESTION:

What is your reason for wanting the position of Fire Captain and what unique qualifications do you possess in relation to this position?

POSSIBLE RESPONSE:

SEE RESPONSES FOR THE PREVIOUS QUESTIONS ABOVE!

QUESTION:

How do you determine or evaluate success?

POSSIBLE RESPONSE:

Respond with your particular method.

QUESTION:

How do you feel that your promotion will **IMPROVE** the __________ Fire Department?

POSSIBLE RESPONSE:

My promotion will help to improve the __________ Fire Department because of my **MOTIVATION** to **IMPROVE MYSELF** and the __________ Fire Department. While performing the duties of Fire Captain I would strive to improve the quality of communication between subordinates, peers, and supervisors. I would make it a point to share information and knowledge with other Fire Captains and aspiring or acting Fire Captains.

ADD OR DELETE ANYTHING APPLICABLE FOR YOU

QUESTION:

Why should **YOU** be the first candidate promoted to the position of Fire Captain on this department?

POSSIBLE RESPONSE:

I should be the **FIRST** candidate promoted to the position of Fire Captain, because I am the **BEST** candidate for the position! I have spent more time and effort preparing myself for this position than any of the other candidates. Securing the position of Fire Captain has been the most important objective in my career for the past several years. I will bring the position **EXPERIENCE, KNOWLEDGE,** and **DEDICATION**. I am ready for the position **RIGHT NOW!** in fact I have been doing the job for several years as an acting Fire Captain.

ADD OR DELETE ANYTHING THAT MAY BE APPROPRIATE !

QUESTION:

What do you think it takes to be successful in the fire service?

POSSIBLE RESPONSE:

Respond with your own ideas.

SEE FOLLOWING RESPONSE!

QUESTION:

What are your **QUALIFICATIONS** for Fire Captain?

POSSIBLE RESPONSE:

My **QUALIFICATIONS** extend from 5 distinct sources:

1. **MOTIVATION.**
2. **TRAINING.**
3. **EDUCATION.**
4. **EXPERIENCE.**
5. **JOB KNOWLEDGE.**

MOTIVATION:

I have a strong desire to improve myself and the _________ Fire Department through involvement in various department projects. GIVE EXAMPLES!

TRAINING:

My training includes reading and very serious study along with inside and outside department drills, and classes, including State Officer Certification. GIVE EXAMPLES!

EDUCATION:

My education includes College degree in Fire Science, along with completion of various Fire Service courses. Currently I am enrolled in various Fire Service courses. GIVE EXAMPLES!

EXPERIENCE:

Give the number of years **YOU** have served in the Fire Service, along with the type of experience you have gained in this period of time. EXAMPLES:

1. Firefighter.
2. Fire Paramedic.
3. Fire Inspector.
4. Arson investigator.
5. Fire Engineer.
6. Acting Fire Captain.
7. Have been responsible for:
   A. Apparatus and equipment.
   B. Subordinates and training of subordinates.
   C. Emergency situations.
   D. Large fires.
   E. All phases of Fire Prevention.
   F. Management and Supervision.
   G. Quarterly task.
   H. Hazardous incidents.
   I. Delegation of task.
   J. Reports and records.
   K. Assignment of personnel and task.
   L. Morale.
   M. Rules and regulations.
   N. Arson investigation.

ADD OR DELETE ANY EXAMPLES THAT ARE APPROPRIATE!

JOB KNOWLEDGE:

I have a thorough knowledge of the requirements for the position of Fire Captain. EXAMPLES:

1. Management-organization.
2. Planning.
4. Leadership-supervision.
5. Delegation-orders.
6. Morale.
7. Reports-records.
8. Training.
9. Firefighting.
10. Fire prevention, target hazards, water supply, and city geography.
11. Department policy and procedures.
12. Department rules and regulations.

ADD OR DELETE ANY EXAMPLES THAT APPLY TO YOU!

QUESTION:

What unique characteristic do you possess that would enhance you in the role as Fire Captain as compared to your competitors?

or

What makes you think that you are more capable of handling this job than any of your competitors?

POSSIBLE RESPONSE:

SEE THE THREE PREVIOUS RESPONSES

QUESTION:

Based on your background and abilities, what in your opinion is your greatest asset that you have to offer the __________ Fire Department?

POSSIBLE RESPONSE:

SEE THE FOUR PREVIOUS RESPONSES

QUESTION:

What recommendations do you have for increasing the **PRODUCTIVITY** of the ________ Fire Department?

POSSIBLE RESPONSE:

My recommendation would be to create **TRAINING PROGRAMS** which would contribute to motivating and stimulating personnel.

ADD WHATEVER YOU FEEL IS APPROPRIATE !

QUESTION:

What do you think **YOUR** greatest problem as a Fire Captain for the ________ Fire Department?

POSSIBLE RESPONSE:

Most likely my greatest problem as Fire Captain for the ________ Fire Department will be in securing **BUDGET APPROVAL** for various equipment.

ADD WHATEVER IS APPROPRIATE FOR YOU !

QUESTION:

What major problem, in the fire service, have you encountered and how did you deal with it?

POSSIBLE RESPONSE:

RESPOND WITH AN APPROPRIATE RESPONSE FOR YOU.

QUESTION:

The position of Fire Captain demands that you accept a great deal of responsibility. In this respect, what is your primary weakness? How are you overcoming this weakness?

POSSIBLE RESPONSE:

Respond with an appropriate response for you.

QUESTION:

What is the **TECHNIQUE** that you use for solving problems?

POSSIBLE RESPONSE:

I solve most problems by making an effort to understand the problem, then researching the solution while allowing **FLEXIBILITY** for the solution. There is more than one solution to almost every problem.

ADD YOUR OWN METHOD OF SOLVING PROBLEMS !

QUESTION:

Describe your management style!

POSSIBLE RESPONSE:

RESPOND WITH AN APPROPRIATE RESPONSE FOR YOU.

QUESTION:

On the __________ Fire Department what do you see as the **MAJOR PROBLEM** at the Fire Captains level? How do you think it could be improved?

POSSIBLE RESPONSE:

At the Fire Captains level on the __________ Fire Department, I consider not being able to get **PERSONNEL INVOLVED** in department projects or motivating them to do their best work as a major problem at this level. I think a solution would be to have subordinates become more involved in training programs that would use their best attributes; such as classroom and field hydraulics for Fire Engineers, rescue procedures for Fire Paramedics, etc.

GIVE APPROPRIATE RESPONSE FOR YOUR DEPARTMENT

QUESTION:

In your opinion, what is the **IMPORTANCE** of the Fire Captains position?

POSSIBLE RESPONSE:

Some of the things that make the position of Fire Captain an important position are:

1. The added **RESPONSIBILITIES** of the knowledge and experience required concerning:
   A. Leadership, organization, planning.
   B. Community relations.
   C. Fire prevention.
   D. Fire suppression.
   E. Fire protection.
   F. Fire hazards and causes.
   G. Building construction.
   H. Codes and ordinances.
   I. Strategy and tactics.
   J. Chemistry.
   K. Safety Practices.
   L. Fire investigation.
   M. Report writing.
   N. Chain of command.
   O. Communication.

ADD OR DELETE ANYTHING YOU FEEL IS APPROPRIATE!

QUESTION:

What additional **TRAINING** do you feel is needed to better prepare future candidates for the position of Fire Captain on the ________ Fire Department?

POSSIBLE RESPONSE:

In order to prepare candidates for the position of Fire Captain, I feel that the ________ Fire Department should have a **CERTIFICATION** program for the position of acting Fire Captain. The program should have guidelines along with a study guide so that each candidate would be able to prepare themselves properly.

ADD WHAT IS APPROPRIATE FOR YOUR DEPARTMENT !

QUESTION:

What part of **SUPERVISORY** work appeals to you the most?

POSSIBLE RESPONSE:

The part of supervisory work that appeals to me the most is **TRAINING**, I get great enjoyment and satisfaction when I am able to help others learn and understand both new and old concepts.

QUESTION:

As **FIRE CAPTAIN**, if it was up to you to make a decision on a situation, would you decide in favor of management or personnel?

POSSIBLE RESPONSE:

In this situation it would depend upon the circumstances. You must have the facts of the case.

QUESTION:

What would you do if you knew that some of the men in your company disliked you?

POSSIBLE RESPONSE:

RESPOND WITH WHAT IS APPROPRIATE FOR YOU.

QUESTION:

If you wanted to win the **ESTEEM** of the men in your company, how would you go about getting it?

POSSIBLE RESPONSE:

RESPOND WITH AN APPROPRIATE ANSWER FOR YOU

QUESTION:

Do you think it would be wise to present more than one solution to your superior when he asks you for recommendations on a particular matter?

POSSIBLE RESPONSE:

RESPOND WITH YOUR APPROPRIATE ANSWER

QUESTION:

What do you see as some of the **PRIMARY PROBLEMS** on the _________ Fire Department?

POSSIBLE RESPONSE:

Some of the **PRIMARY PROBLEMS** on the department are:
1. TRAINING.
2. BUDGETING.
3. EQUIPMENT.
4. MAINTENANCE of EQUIPMENT.
5. MORALE.

ADD OR DELETE ANY APPROPRIATE RESPONSE !

QUESTION:

In your opinion what is the greatest **INNOVATION** in the Fire Service in the past few years?

POSSIBLE RESPONSE:

In my opinion the greatest **INNOVATION** in the past few years is:
HAZARDOUS MATERIAL SQUADS.
TELL THE BOARD WHAT YOU FEEL STRONGLY ABOUT !

QUESTION:

What do you feel is the **PROMINENT** problem facing the Fire Service today?

POSSIBLE RESPONSE:

I feel that **HAZARDOUS MATERIALS** in use, storage and transportation is the most **PROMINENT** problem.

ADD YOUR THOUGHTS !

QUESTION:

As a new Fire Captain, what **NEW INNOVATIVE** ideas can you introduce to the _________ Fire Department.

POSSIBLE RESPONSE:

An **INNOVATIVE** idea that I could contribute to the _________ Fire Department would be the compilation of a Fire Captains **STUDY GUIDE**.

ADD APPROPRIATE RESPONSE FOR YOU !

QUESTION:

How would/could you **IMPROVE** the _________ Fire Department?

POSSIBLE RESPONSE:

Give an appropriate response for you and your particular Fire Department. **SEE ABOVE RESPONSE!**

QUESTION:

What two or three **ACCOMPLISHMENTS** have given you the most satisfaction? Why?

POSSIBLE RESPONSE:

RESPOND WITH WHATEVER IS APPROPRIATE FOR YOU.

QUESTION:

How would you **CHANGE** the _______ Fire Department?

POSSIBLE RESPONSE:

Some **CHANGES** that I believe would **IMPROVE** the _______ Fire Department would be:

1. Make **OUTSIDE TRAINING** more readily available for personnel.
2. Involve a greater number of personnel in budgeting for training needs, tools and equipment, ETC.

ADD YOUR OWN THOUGHTS !

QUESTION:

What type of program(s) are needed in the _______ Fire Department?

POSSIBLE RESPONSE:

Give an appropriate response for your particular Fire Department. **SEE PREVIOUS RESPONSE!**

QUESTION:

What **CHANGES** would you make to increase the efficiency of the _______ Fire Department?

POSSIBLE RESPONSE:

Give an appropriate response for your particular Fire Department. **SEE PREVIOUS RESPONSE!**

QUESTION:

What is your **FIRST GOAL** after you are promoted to Fire Captain?

POSSIBLE RESPONSE:

My **FIRST GOAL** after being promoted is to be the best Fire Captain that I can possibly be.

GIVE YOUR APPROPRIATE RESPONSE !

QUESTION:

What are your immediate and ultimate **GOALS** in the Fire Service?

POSSIBLE RESPONSE:

My **IMMEDIATE** goal is to be the best Fire Captain that I can possibly be. My **ULTIMATE** goal in the Fire Service is to contribute as much to the Fire Service that my abilities will allow.

ADD ANYTHING THAT MAY APPLY TO YOU !

QUESTION:

What are your long range **GOALS** in the fire service?
How do you plan to achieve these goals?

POSSIBLE RESPONSE:

Respond with whatever is appropriate for you. **SEE PREVIOUS TWO RESPONSES!**

QUESTION:

What are your long range goals and short range goals and objectives and why did you establish these goals and how are you preparing yourself to achieve them?

POSSIBLE RESPONSE:

Respond with whatever is appropriate for you, such as continuing to gain experience and knowledge through training and involvement. **SEE PREVIOUS THREE RESPONSES!**

QUESTION:

Do your peers **RESPECT** you in your present position with the __________ Fire Department?

POSSIBLE RESPONSE:

YES. RESPOND WITH WHATEVER IS APPROPRIATE FOR YOU.

QUESTION:

What specific goals, other than those related to the fire service, have you established for yourself for the next ten (10) years?

POSSIBLE RESPONSE:

RESPOND WITH WHAT IS APPROPRIATE FOR YOU.

QUESTION:

What do you see yourself doing in the fire service five (5) years from now?

POSSIBLE RESPONSE:

Respond with whatever you feel is appropriate for you. Example: five years from now I see myself continuing to gain experience and knowledge through training and involvement in the Fire Service.

QUESTION:

What two or three things are the most important to you in your Fire Service **CAREER**?

POSSIBLE RESPONSE:

The three most important things for me in my **CAREER** are:

1. Job satisfaction.
2. Involvement.
3. Security. **INSERT YOUR OWN IDEAS**

QUESTION:

Will you, or do you, expect any problems in becoming the supervisor of your close friends?

POSSIBLE RESPONSE:

No, my peers have respect for me and the position of Fire Captain.

ADD WHATEVER IS APPROPRIATE FOR YOU

QUESTION:

Describe your most **REWARDING** job experience in the Fire Service.

POSSIBLE RESPONSE:

Respond with your most rewarding job experience in the Fire Service. EXAMPLE:

During the period of my career that I was a Paramedic, I handled an incident where the brother of my Captain was in full arrest and our crew was able to completely revive the man at the scene of the incident.

QUESTION:

What is the biggest **MISTAKE** that you have made since coming into the Fire Service?

POSSIBLE RESPONSE:

My **BIGGEST MISTAKE** since coming into the Fire Service would be that I did **NOT** get **INVOLVED** in the **PROMOTIONAL PROCESS** earlier in my career.

GIVE YOUR APPROPRIATE RESPONSE !

QUESTION:

What **FUTURE PROBLEMS** do you anticipate in the Fire Service?

POSSIBLE RESPONSE:

Some of the **FUTURE PROBLEMS** that I would anticipate would be concerned with:

1. HAZARDOUS MATERIALS.
2. BUDGETING.
3. APPROPRIATE MANPOWER.

ADD YOUR THOUGHTS !

QUESTION:

What are the major **ISSUES** facing the __________ Fire Department today?

POSSIBLE RESPONSE:

SEE RESPONSE TO THE PREVIOUS QUESTION!

QUESTION:

What type of **LEADER** are you?

POSSIBLE RESPONSE:

I am a combination of **AUTOCRATIC** and **DEMOCRATIC** with the idea of teamwork depending on the situation. During an emergency situation I would be totally autocratic.

QUESTION:

What is your philosophy of **"APPARATUS MAINTENANCE"**?

POSSIBLE RESPONSE:

APPARATUS MAINTENANCE is very important for the:

1. **SAFETY** of personnel and citizens.
2. **QUALITY** and **DURATION** of apparatus use.

ADD SOME OF YOUR OWN THOUGHTS OR IDEAS !

QUESTION:

How **AGGRESSIVE** do you feel an Apparatus Driver should be while driving **CODE 3** ?

POSSIBLE RESPONSE:

I feel that the Apparatus Driver should drive **COMPLETELY DEFENSIVE**, always willing to give up the right of way!

ADD YOUR OWN THOUGHTS !

QUESTION:

If you were assigned to a special project to help **CUT** the **DEPARTMENTS BUDGET** by **10%**. What recommendations would you make for accomplishing this assignment?

POSSIBLE RESPONSE:

MAKE AN APPROPRIATE RESPONSE WHICH PERTAINS TO YOUR FIRE DEPARTMENT !

QUESTION:

Of all the **COURSES** that you have taken in College, special seminars; which do you feel has been the **MOST BENEFICIAL** to you?

POSSIBLE RESPONSE:

The class which has been the **MOST BENEFICIAL** for me was through the States Fire Officers Certification Program: **FIRE ADMINISTRATION AND MANAGEMENT**. (be prepared to tell the board why you feel that any particular training has been beneficial to you!)

ADD OR DELETE ANYTHING THAT IS APPLICABLE TO YOU !

QUESTION:

What part of your formal **EDUCATION,** obtained **BEFORE** you were employed by the ________ Fire Department, do you think will help you most as a Fire Captain?

POSSIBLE RESPONSE:

Prior to becoming a Firefighter, the class of **PUBLIC SPEAKING,** which enhanced my ability to communicate and interrelate with other people.

OR:

Prior to becoming a Firefighter I attended the **MANAGEMENT** and **SUPERVISION** course within the state officers training curriculum.

RESPOND WITH WHATEVER IS APPROPRIATE FOR YOU

QUESTION:

Which one or more of the **COURSES** taken recently do you think would help you the most as an officer?

POSSIBLE RESPONSE:

The course that has helped me the most is the **MANAGEMENT** and **SUPERVISION** course within the state fire officers certification curriculum.

QUESTION:

Give an example of where you have displayed **LOYALTY** to the ________ Fire Department.

POSSIBLE RESPONSE:

I have displayed **LOYALTY** to the ______ Fire Department through my **STRONG MOTIVATION** to **IMPROVE MYSELF** and the **DEPARTMENT!**

GIVE SOME EXAMPLES THAT APPLY TO YOU !

QUESTION:

What is your opinion of **WOMEN** in the Fire Service?

POSSIBLE RESPONSE:

GIVE YOUR OWN APPROPRIATE RESPONSE !

QUESTION:

What type of **COMMUNITY ACTIVITIES** have or are you involved in?

POSSIBLE RESPONSE:

I have been involved in such activities as:
1. FIRE SERVICE DAY.
2. Accumulated funds for the HOSPITAL BURN WARD.
MAKE AN APPROPRIATE RESPONSE THAT PERTAINS TO YOU!

QUESTION:

What do you feel is the **PRIMARY VALUE** of conducting **FIRE PREVENTION** at the **COMPANY LEVEL?**

POSSIBLE RESPONSE:

Conducting **FIRE PREVENTION** at the **COMPANY LEVEL** will give personnel gained **FAMILIARIZATION** of occupancies and target hazards along with **GOOD PUBLIC RELATIONS**.
ADD YOUR OWN THOUGHTS !

QUESTION:

What's your opinion of **AFFIRMATIVE ACTION PROGRAMS?**

POSSIBLE RESPONSE:

GIVE YOUR OWN APPROPRIATE RESPONSE !

QUESTION:

Explain the areas that you feel **SENIORITY** should be used in the fire service.

POSSIBLE RESPONSE:

GIVE YOUR OWN APPROPRIATE RESPONSE!

QUESTION:

What is your opinion of the testing procedure used for this promotional exam?

POSSIBLE RESPONSE:

EXAMPLE:
I think that the testing procedure (Assessment Center) for this exam has covered all the essentials for the position of Fire Captain along with allowing enough time for the assessors to get to know each of the candidates so as to make an evaluation of each candidate. Give an appropriate response for your feelings of your particular testing procedure.

QUESTION:

What **BOOKS** and **MAGAZINES** have you **READ** in preparation for this promotional exam?

POSSIBLE RESPONSE:

Some examples of books and magazines that I have read are:

1. The Fire Chiefs Handbook.
2. Management of Fire Service Operations.
3. Fire Attack.
4. Fire Engineer Written and Oral Exam Study Guides.
5. Fire Captain Written and Oral Exam Study Guides.
6. Engine Company 82. & Thirty Years On The Line.
7. IFSTA MANUALS
8. Incident Command System
9. Firehouse Magazine, Firefighters News, Fire Chief Magazine, and Fire Engineering.

MAKE AN APPROPRIATE LIST WHICH PERTAINS TO YOU !

QUESTION:

Why are **YOU** the **BEST PERSON** for the position ?
THIS IS A CHANCE TO SELL YOURSELF, UTILIZE IT!

POSSIBLE RESPONSE:

I am the **BEST PERSON** for the position because I have spent more **TIME** and **EFFORT** preparing myself for the position. Obtaining the position of Fire Captain has been the **MOST IMPORTANT** priority in my career for the past several years. I will bring the position **EXPERIENCE, KNOWLEDGE,** and **DEDICATION**. I am ready to do the job **NOW!** in fact I have been doing the job several years as an acting Captain.
ADD ANYTHING THAT MAY APPLY TO YOU !

QUESTION:

MANAGERIALLY speaking, what **BAD HABITS** or practices that you now have would you like to change? everyone has faults, tell us what yours are and how you are going to change them!
THIS IS A GOOD TIME TO TRY AND TURN A NEGATIVE INTO A POSITIVE!

POSSIBLE RESPONSE:

Sometimes I drive myself quite hard and I'm rather exacting and there are times that I might expect too much of others.

I am changing these problems now, just by being aware of the problems. I am able to correct the fault with self awareness as it is happening.

QUESTION:

Is there any area that has not been touched on, or is there anything that you would like to **ADD?**

POSSIBLE RESPONSE:

YES!, I have been preparing myself for the position of Fire Captain for several years. Obtaining the position of Fire Captain has been the **NUMBER ONE PRIORITY** while I have been preparing myself!

I HAVE BEEN PREPARING MYSELF THROUGH:

1. Reading and serious study.
2. Training and education.
3. Hands-on experience as an acting Fire Captain.

I AM READY TO DO THE JOB NOW, AND WILL BRING THE POSITION:

1. Knowledge.
2. Experience.
3. Dedication.

add whatever you feel is appropriate at this time. This is the time the board will give you the chance to "blow your horn" about yourself !

BE CALM, MOST OF ALL BE YOURSELF, DON'T "FAKE IT"!

SECTION 3

GENERAL KNOWLEDGE

FIRE SERVICE

QUESTION:

What are some of the **CURRENT EVENTS** that are having an effect on the fire service and what can the fire service do to solve these problems?

POSSIBLE RESPONSE:

Read all the trade magazines and stay current with issues and problems facing the fire service!

EXAMPLES:

1. High rise problems.
2. Haz Mat problems.
3. Transportation problems.ETC.

QUESTION:

What is **FIRE SCOPE**?

POSSIBLE RESPONSE:

FIRE SCOPE is the Firefighting Resources of California Organized for Potential Emergencies.

FIRE SCIENCE

QUESTION:

What is **CONVECTION**?

POSSIBLE RESPONSE:

CONVECTION is the transfer of heat by the movement of air or liquid.

QUESTION:

What is **RADIATION**?

POSSIBLE RESPONSE:

RADIATION is heat in the form of rays that travels through space until it reaches an opaque object.

QUESTION:

Define **FLAMMABLE LIMITS**?

POSSIBLE RESPONSE:

FLAMMABLE LIMITS: the maximum or minimum concentration of combustible gas and air mixture that will ignite.

QUESTION:

What is **IGNITION TEMPERATURE**?

POSSIBLE RESPONSE:

IGNITION TEMPERATURE is the minimum temperature to which a substance must be heated in order to cause it to burn.

QUESTION:

Explain the **FIRE TRIANGLE**.

POSSIBLE RESPONSE:

The **FIRE TRIANGLE** is a three sided figure representing the three of the four factors necessary for combustion = oxygen, heat, and fuel. The **FIRE TRIANGLE** has been replaced by the **TETRAHEDRON**.

QUESTION:

Explain the **FIRE TETRAHEDRON**.

POSSIBLE RESPONSE:

The **FIRE TETRAHEDRON** is a figure of four sides representing the four elements required for fire = fuel, heat, oxygen, and uninhibited chain reactions. Each side is adjacent with the other three sides.

QUESTION:

Can you explain the **LAW OF HEAT FLOW?**

POSSIBLE RESPONSE:

The **LAW OF HEAT FLOW** refers to the fact that heat tends to flow from **HOT TO COLD**.

QUESTION:

What is the **LAW OF SPECIFIC HEAT**?

POSSIBLE RESPONSE:

SPECIFIC HEAT refers to the heat capacity of a substance as compared to the heat capacity of water.

QUESTION:

Briefly explain **FLASHOVER**.

POSSIBLE RESPONSE:

FLASHOVER is when a fire is to the point in which all surfaces and objects are heated to their ignition temperature and flame breaks out almost instantly over the entire surface.

QUESTION:

What is **VAPOR DENSITY**?
How does it relate to fire prevention purposes?

POSSIBLE RESPONSE:

VAPOR DENSITY is the weight per unit volume of a pure gas or vapor. For fire prevention purposes, **VAPOR DENSITY** is reported in terms of the ratio of the relative weight of a volume of vapor to the weight of an equal volume of air under the same conditions of temperature and pressure.

QUESTION:

What are the fundamental rules that govern **FRICTION LOSS** in hose?

POSSIBLE RESPONSE:

Fundamental rules that govern **FRICTION LOSS** are:

1. Varies with the relative quality of the hose.
2. Varies directly with the length of the hose.
3. Varies with the square of the water velocity.
4. Varies inversely as the fifth power of the hose diameter, for a given velocity.
5. Is independent of the pressure, for a given velocity.

QUESTION:

What are the six basic principles that apply to **FLUID PRESSURE**?

POSSIBLE RESPONSE:

The six principles that apply to **FLUID PRESSURE:**

1. Liquid pressure is exerted in a perpendicular direction to any surface on which it acts.
2. At any given point beneath the surface of a liquid, the pressure is the sam in all directions, downward, upward, and sideways.
3. Pressure applied to a confined liquid from without is transmitted in all directions without reduction in intensity.
4. The pressure of a liquid in an open vessel is proportional to the density of the liquid.
5. The pressure of a liquid in an open vessel is proportional to the depth of the liquid.
6. Liquid pressure on the bottom of a vessel is unaffected by the size or shape of the vessel.

WATER SUPPLY

QUESTION:

What are some of the **STRONG POINTS** of the **WATER SYSTEM** servicing this City?

POSSIBLE RESPONSE:

STRONG POINTS of this Cities **WATER SYSTEM** are:

1. System is **NOT** an **ISOLATED** system, but part of a large system with **GREATER RELIABILITY** and **FLEXIBILITY**.
2. **GRID SYSTEM**, which is an arrangement of water mains with lateral feeders for improved distribution.
3. GIVE EXAMPLES of large amounts of **WATER STORAGE** within the your City limits.
4. GIVE EXAMPLES of **BOOSTER PUMPS** within your City limits.
5. GIVE EXAMPLES of some of the **LARGER MAINS** within your City limits.
6. GIVE EXAMPLES of some of the **STRONG POINTS** that pertain to your Cities **WATER SYSTEM**.

QUESTION:

What are some of the **WEAK POINTS** of the **WATER SYSTEM** servicing the City that you work in?

POSSIBLE RESPONSE:

Some examples of **WEAK POINTS** in this Cities **WATER SYSTEM:**

1. GIVE location of **DEAD END MAINS**.
2. GIVE location of **SMALL DIAMETER MAINS**.
3. GIVE location of **LOW GPM HYDRANTS**.
4. GIVE location of **TARGET HAZARDS** with any of 1-3.
5. GIVE EXAMPLES of some of the **WEAK POINTS** that pertain to your Cities **WATER SYSTEM**.

QUESTION:

How much water is **STORED** within your Cities boundaries?

POSSIBLE RESPONSE:

GIVE APPROPRIATE ANSWER FOR YOUR CITIES WATER SYSTEM

QUESTION:

What is the **REQUIRED FIRE FLOW** for your City?

POSSIBLE RESPONSE:

GIVE THE APPROPRIATE ANSWER FOR YOUR FIRE DEPT.

QUESTION:

Where does your City receive its **WATER SUPPLY** from?

POSSIBLE RESPONSE:

1. Rivers or lakes.
2. Artisan wells.
3. Storage tanks.
4. Other water districts.

GIVE EXAMPLES FOR YOUR CITIES WATER SYSTEM

QUESTION:

What is the **AVERAGE STATIC HYDRANT PRESSURE** in your City?

POSSIBLE RESPONSE:

GIVE PROPER ANSWER FOR CITIES HYDRANT SYSTEM

QUESTION:

What is the **ISO RATING** for your Cities **WATER SYSTEM**?

POSSIBLE RESPONSE:

GIVE APPROPRIATE ANSWER FOR CITIES WATER SYSTEM

QUESTION:

Where are your Cities **WATER STORAGE** and **BOOSTER TANKS** located?

POSSIBLE RESPONSE:

GIVE APPROPRIATE ANSWER FOR YOUR FIRE DEPT.

ISO-AND CITY INFORMATION

QUESTION:

What is the size and population of your City?

POSSIBLE RESPONSE:

GIVE APPROPRIATE FIGURES FOR YOUR CITY

QUESTION:

What class **ISO RATING** is the __________ Fire Department?

POSSIBLE RESPONSE:

GIVE THE ISO RATING THAT YOUR DEPARTMENT HAS

QUESTION:

When was the last **ISO SURVEY** conducted in your City?

POSSIBLE RESPONSE:

GIVE THE DATE OF THE LAST **ISO SURVEY** WAS CONDUCTED

QUESTION:

Who is the **MAYOR** and who are the **CITY COUNCIL** members for your City?

POSSIBLE RESPONSE:

GIVE THE APPROPRIATE NAMES FOR YOUR CITY

QUESTION:

Explain any **MUTUAL** or **AUTOMATIC AID** agreements that your Cities Fire Department may have with others.

POSSIBLE RESPONSE:

GIVE APPROPRIATE ANSWER FOR YOUR CITY

FIRE APPARATUS AND EQUIPMENT

IT IS NORMALLY ASSUMED THAT THE QUESTIONS IN THIS SECTION HAVE BEEN COVERED IN OTHER PROMOTIONAL EXAMS, BUT THEY ARE ASKED DURING FIRE OFFICER EXAMS ON OCCASION!

QUESTION:

What constitutes a **TRIPLE COMBINATION** apparatus?

POSSIBLE RESPONSE:

A **TRIPLE COMBINATION** apparatus is an apparatus with a **PUMP, WATER TANK**, and **HOSE**.

QUESTION:

What constitutes a **QUADRUPLE COMBINATION** apparatus?

POSSIBLE RESPONSE:

A **QUADRUPLE COMBINATION** apparatus is an apparatus with a **PUMP, WATER TANK, HOSE,** and **GROUND LADDERS**.

QUESTION:

What constitutes a **"QUINT"** ? (quintuple combination apparatus)

POSSIBLE RESPONSE:

A **"QUINT"** is an apparatus with a **PUMP, WATER TANK, HOSE, GROUND LADDERS**, and an **AERIAL LADDER**.

QUESTION:

What is the **PRINCIPLE** of the **CENTRIFUGAL PUMP?**

POSSIBLE RESPONSE:

The **PRINCIPLE** of the **CENTRIFUGAL PUMP** is the tendency of revolving body to fly outward from the center of rotation.

QUESTION:

What is the difference between a **SINGLE STAGE PUMP** and a **TWO STAGE PUMP**, and how do they work?

POSSIBLE RESPONSE:

A **SINGLE STAGE PUMP** has only one impeller. The **TWO STAGE PUMP** has two impellers which allows it to pump in **VOLUME** (parallel, capacity) for maximum GPM or in **SERIES** (pressure) which allows for higher pressures. Accomplished by use of **TRANSFER VALVE**.

QUESTION:

Explain how a **PRESSURE RELIEF VALVE** on your departments fire apparatus works.

POSSIBLE RESPONSE:

PRESSURE RELIEF VALVES prevent pressures from going above the desired setting by bypassing water from the discharge outlet back to the suction inlet.

The **PRESSURE RELIEF VALVE** is set at the desired pressure on the pump panel, while water is flowing. If the pressure from the discharge side of the pump becomes greater than what the **RELIEF VALVE** is set at, a spring in the pilot valve compresses, which in turn moves a needle valve, which permits water to flow back to the intake side of the pump. This action reduces the pressure in the main valve tube.

For your departments apparatus, describe how their particular **RELIEF VALVES** function. If your departments apparatus use **GOVERNORS**, describe how they operate.

QUESTION:

What is the primary purpose of **PRESSURE CONTROL DEVICES**? Name three different types of devices.

POSSIBLE RESPONSE:

The primary purpose of **PRESSURE CONTROL DEVICES** is to protect men at the nozzles from a dangerous pressure rise. (a secondary purpose is to protect the hose and pumps).

1. AUTOMATIC RELIEF VALVE.
2. PRESSURE OPERATED GOVERNOR.
3. PRESSURE REDUCING VALVES.

QUESTION:

Explain how the **GOVERNORS** on your departments apparatus pumps operate.

POSSIBLE RESPONSE:

GOVERNORS function by controlling the discharge pressure by regulating the speed of the engine.

The **GOVERNOR** assembly is an integral part of the throttle linkage and operates by balancing pump pressure against a spring-loaded piston within a cylinder (flow control unit and valves). As pump pressure is increased the governor spring is compressed which lengthens the governor assembly. As long as the pump discharge pressure and the governor spring pressure are balanced, the throttle butterfly valve is held at the throttle setting. If a nozzle is shut down, pump pressure will increase, which causes greater pressure, which will compress the governor spring, which will lengthen the governor assembly and close the throttle. If the pump pressure decreases, the governor spring pressure will be greater than the water pressure, which will shorten the governor assembly and open the throttle.

QUESTION:

What are the three different methods of controlling **PUMP PRESSURES?**

POSSIBLE RESPONSE:

1. **AUTOMATIC RELIEF VALVE** that opens a bypass between the suction and discharge sides of pump.
2. **PRESSURE-OPERATED GOVERNOR** to control speed setting of the throttle on the engine.
3. **PRESSURE REDUCING VALVES** on each discharge gate.

QUESTION:

How does **ALTITUDE** affect the power developed by gasoline engines?

POSSIBLE RESPONSE:

Power developed by gasoline engines decreases **3 1/2%** per 1,000 foot of **ALTITUDE** above sea level.

QUESTION:

At what altitude does a **TURBOCHARGED DIESEL ENGINE** start to have a power loss and at what rate?
Also **NORMALLY ASPIRATED DIESEL ENGINES**.

POSSIBLE RESPONSE:

The **TURBOCHARGED DIESEL ENGINE** starts to lose power at the 4000 foot level at a rate of 2% per 1000 foot of elevation in excess of 4000 feet.

NORMALLY ASPIRATED DIESEL ENGINES start to lose power at altitudes of 1000 feet above sea level at a rate of 3% per 1000 feet of altitude.

QUESTION:

What do you think of the practice of **STARTING** fire apparatus engines at the change of each shift and why?

POSSIBLE RESPONSE:

It is unnecessary and of no practical value to **START** modern fire apparatus at the change of each shift because:

1. It causes needless ware and abuse.
2. Does not prove that the engine will start the next time.
3. A large portion of engine wear takes place during first few seconds of starting.
4. When engine is cold, water and fuel condense on working parts, causing corrosion of valve lifters, springs, and other internal parts.
5. Water vapor mixes with Carbon and dirt particles in the oil, forming emulsion to sludge.

QUESTION:

What is the proper weight to horsepower **RATIO** in a good fire truck?

POSSIBLE RESPONSE:

The proper weight to horsepower **RATIO** of a good fire truck is **88 LBS** to the horsepower. Raise
550 LBS 1 foot in one second, or 33,000 LBS 1 foot
in one minute.

QUESTION:

Explain to us how a **CENTRIFUGAL PUMP** works.

POSSIBLE RESPONSE:

A **CENTRIFUGAL PUMP** develops pressure in the fluid (water) being pumped by creating kinetic (moving) energy to the liquid by means of the rotation of the impellers. The water enters the impeller near its center (inlet eye) and is whirled by the impeller vanes; this action develops centrifugal force which urges the water outward through the volute into the discharge of the pump.

QUESTION:

On specifications, what is the **SINGLE** most important decision concerning the **CHASSIS**?

POSSIBLE RESPONSE:

The **SINGLE** most important decision concerning the **CHASSIS** is the selection of the engine.

QUESTION:

How is a pumps **RATED CAPACITY** determined?

POSSIBLE RESPONSE:

Pumps **RATED CAPACITY** is determined by **TESTING**.

QUESTION:

What is the **MAXIMUM NO-LOAD GOVERNED SPEED** of an engine?

POSSIBLE RESPONSE:

The **MAXIMUM NO-LOAD GOVERNED SPEED** is established by the engine manufacture as a safe limit of engine speed. The governor will prevent the engine from exceeding the safe speed. Most engine manufactures allow a plus tolerance of **2%** for **MAXIMUM NO-LOAD GOVERNED SPEED.**

QUESTION:

What are the affects of water temperatures over **85 DEGREES F.** and at **95 DEGREES F**, in a pump?

POSSIBLE RESPONSE:

Affects of water temperature over **85 DEGREES F** in a pump would be problems in priming and obtaining capacity discharge. At **95 DEGREES F** lift capability is greatly reduced and only a small portion of rated capacity can be delivered. Water flow will be erratic. As water is heated, its tendency is to give off vapors more rapidly.

QUESTION:

Explain **NET PUMP PRESSURE** while pumping from a hydrant and from draft.

POSSIBLE RESPONSE:

NET PUMP PRESSURE is the pressure actually developed in the pump.
NET PUMP PRESSURE while pumping from a hydrant is the difference between the intake pressure and the discharge pressure.

NET PUMP PRESSURE while pumping at draft is the sum of the static suction lift, suction hose friction loss, losses in the strainer and the velocity head.

These are all added to the discharge pressure to give the total **NET PUMP PRESSURE**.

QUESTION:

What is the **MAIN ACTION** of a fire pump?

POSSIBLE RESPONSE:

The **MAIN ACTION** of any fire pump is to **ADD PRESSURE** to the water.

QUESTION:

What are your departments policy and procedures for **MAINTENANCE** of apparatus?

POSSIBLE RESPONSE:

Give appropriate response for such things as:

1. Daily check-off sheet.
2. Weekly check-off sheet.
3. Monthly service.
4. 6 month service.
5. Annual service.

ANYTHING THAT IS APPROPRIATE FOR YOUR DEPARTMENT

QUESTION:

What is the **MAXIMUM PRESSURE** that you can safely pump at with the pumping apparatus on your department?

POSSIBLE RESPONSE:

Give the appropriate standard for your department.

QUESTION:

How can you tell the **CAPACITY** of a fire pumper?

POSSIBLE RESPONSE:

You can tell the **CAPACITY** of a fire pumper by multiplying the number of 2 1/2" discharge gates by 250 GPM. You can also find out by testing.

QUESTION:

What is the total **PUMPING CAPACITY** of all the fire pumpers at the station that you are assigned? What is the total **PUMPING CAPACITY** of all the fire pumpers in service for your department?

POSSIBLE RESPONSE:

Give the appropriate **PUMPING CAPACITIES** for the station that you are assigned and for your department.

QUESTION:

When should a fire pumper have its **SERVICE TEST**?

POSSIBLE RESPONSE:

Fire pumpers should have **SERVICE TEST** annually and after major repairs.

QUESTION:

What are the requirements for the **SERVICE TEST** of fire pumpers?

POSSIBLE RESPONSE:

SERVICE TEST AT DRAFT:
1. 100% volume @ 150 PSI for 20 minutes.
2. 100% volume @ 165 PSI for 5 minutes.(not req.)
3. 70% volume @ 200 PSI for 10 minutes.
4. 50% volume @ 250 PSI for 10 minutes.

When testing, the vertical distance from the surface of the water to the center of the pump suction inlet should not exceed 10 feet.

During the **SERVICE TEST**:
1. Use no more than 20 feet of hard suction with the strainer at least 2 feet below the surface of the water.
2. Water should be at least 4 feet deep.
3. Strainer should be at least 2 feet from the bottom of the water source.

QUESTION:

What are the requirements for the **CERTIFICATION TEST** of fire pumpers?

POSSIBLE RESPONSE:

Same as service test **EXCEPT**:
1. 100% volume @ 150 PSI for **2 HOURS**.
2. 100% volume @ 165 PSI for 10 minutes.(spurt)
3. 70% volume @ 200 PSI for **30 MINUTES**.
4. 50% volume @ 250 PSI for **30 MINUTES**.

QUESTION:

In a centrifugal pump, what happens to the **FLOW** of water through the pump, if the discharge gates are suddenly closed?

POSSIBLE RESPONSE:

Sudden shutting of the discharge gates of a centrifugal pump will stop the **FLOW** of water through the pump.

QUESTION:

What are the requirements for the **DELIVERY TEST, (ACCEPTANCE TEST)** at draft?

POSSIBLE RESPONSE:

The **DELIVERY TEST** requirements are the same as the certification test, **PLUS**:

The **ROAD TEST**:
1. The driving performance.
2. Carrying capacity.
3. Cooling system.
4. Suspension.
5. Braking system.

The **ROAD TEST** should be conducted by joint supervision of representatives from the manufacture and the fire department. The **CERTIFICATION TEST** is considered the most important test. It is the baseline for later comparisons with the service test.

QUESTION:

What **HOSE** does NFPA recommend that each engine company carry?

POSSIBLE RESPONSE:

Each engine company should carry:
1. 1200 feet of 2 1/2 inch hose or larger.
2. 400 feet of 1 1/2 inch hose.
3. 200 feet of 1 inch hose.

QUESTION:

What are the requirements for the annual service test of **COTTON DOUBLE JACKET HOSE**?

POSSIBLE RESPONSE:

COTTON DOUBLE JACKET HOSE should be tested annually:
1. At 250 PSI.
2. For 5 minutes.
3. In 300 foot lengths.

QUESTION:

What is the greatest advantage that **COTTON DOUBLE JACKET HOSE** has over single jacket?

POSSIBLE RESPONSE:

The greatest advantage that **COTTON DOUBLE JACKET HOSE** has over single jacket hose is **DURABILITY**.

QUESTION:

What is a **HYDROMETER** scale calibrated to read?

POSSIBLE RESPONSE:

HYDROMETER reads battery **SPECIFIC GRAVITY.**

QUESTION:

How do **DRY CHEMICAL EXTINGUISHERS** extinguish?

POSSIBLE RESPONSE:

DRY CHEMICAL EXTINGUISHERS extinguish by:

1. Smothering.
2. Cooling.
3. Radiation shielding.
4. Chain breaking. (a reaction in the flame may be the principle cause of extinguishment).

QUESTION:

What is **SPECIFIC GRAVITY?**

POSSIBLE RESPONSE:

SPECIFIC GRAVITY is the weight or mass of a given volume of matter at a designated temperature, as compared to that of an equal volume of some other matter. **SPECIFIC GRAVITY** is a ratio of weight to volume.

QUESTION:

Explain **SPECIFIC GRAVITY** in conjunction with the use of a hydrometer, for checking batteries.

POSSIBLE RESPONSE:

SPECIFIC GRAVITY or hydrometer reading of a fully charged battery, should not be less than **1.260** (1.260 - 1.280). These readings are for electrolyte standards at temperatures of 80 degrees F. The **SPECIFIC GRAVITY** varies with the quantity of acid in the solution along with the temperature. As the temperature drops the electrolytes contract and **SPECIFIC GRAVITY** increases: +.004 or +4 points for each 10 degrees F above 80 degrees F and -.004 or -4 points for each 10 degrees F below 80 degrees F.

If **SPECIFIC GRAVITY** varies more than **25** gravity points between cells, replace the battery.

QUESTION:

What are the elements of battery **ELECTROLYTE**?

POSSIBLE RESPONSE:

Battery **ELECTROLYTE** is **SULFURIC ACID** and **WATER**.

QUESTION:

When should fire apparatus have **AIR BRAKES**?

POSSIBLE RESPONSE:

AIR BRAKES on fire apparatus if over 25,000 GVWR.

QUESTION:

What **GASES** are released from a storage battery?

POSSIBLE RESPONSE:

The **GASES** that are released from a storage battery are: **HYDROGEN** and **OXYGEN**.

QUESTION:

Explain what you know about the **FOAM** that your department uses.

POSSIBLE RESPONSE:

The **FOAM** that my department uses is:

AFFF (**A**queous **F**ilm-**F**orming **F**oam), a **FOAM** concentrate based on fluorinated surfactant plus **FOAM** stabilizers and diluted with water to form a 3% or a 6% solution. The **FOAM** excludes air or Oxygen and develops an aqueous film capable of suppressing fire.
3% solution (3% **FOAM** and 97% water) used for Hydrocarbons or petroleum products.

6% solution (6% **FOAM** and 94% water) used for Polar-Solvents or water soluble products.

RESPOND WITH THE APPROPRIATE DESCRIPTION OF THE FOAM THAT YOUR DEPARTMENT USES.

QUESTION:

Describe **CONVENTIONAL FOAM**:

POSSIBLE RESPONSE:

CONVENTIONAL FOAM is formed by the reaction of alkaline salt solution in acid salt solution in the presence of a foam stabilizing agent, and mechanical or air foam formed by turbulent mixing of air with water containing foam forming agents.

QUESTION:

What are some of the substances that will **BREAKDOWN ORDINARY FOAM**?

POSSIBLE RESPONSE:

ORDINARY FOAM is **BROKEN DOWN** by:
1. Common alcohols.
2. Aldehydes.
3. Ethers.

QUESTION:

What is **LIGHT WATER** (fluorinated surfactant) used for:

POSSIBLE RESPONSE:

LIGHT WATER is useful in obtaining quick knockdown of flammable liquid fires, and in providing a vapor sealing effect for reducing subsequent flashover of fuel vapors exposed to lingering open flames.

QUESTION:

What is **ALCOHOL FOAM** recommended for?

POSSIBLE RESPONSE:

ALCOHOL FOAM is recommended for all water soluble flammable liquids, except for those that are only very slightly soluble.

QUESTION:

What different types of **HOSE STREAMS** are used in the fire service?

POSSIBLE RESPONSE:

HOSE STREAMS that the fire service uses, are:
1. Fog streams.(master stream and hand line)
2. Straight streams.(master stream and hand line)

QUESTION:

How much **WATER** can you get from a single 2 1/2" hose line?

POSSIBLE RESPONSE:

You can get from **250 GPM** to **300 GPM** through a single 2 1/2" hose line.

QUESTION:

What are the advantages and disadvantages of **VISCOUS WATER** (thickened water)?

POSSIBLE RESPONSE:

The advantages of **VISCOUS WATER** are:

1. Sticks to burning fuel.
2. Spreads itself in continuous coating.
3. Thicker than plain water.
4. Absorbs more heat.
5. Projects longer and higher straight streams.
6. Seals fuel from Oxygen after drying.
7. Resist wind drift. (as from aircraft in forest fires).

The disadvantages of **VISCOUS WATER ARE**:

1. Does not penetrate fuel as well.
2. Increases friction loss in hose and pipes.
3. Increases water droplet size.
4. Makes working areas slippery.

QUESTION:

What is considered to be the maximum **PENETRATION ANGLE** for a fire stream in the fire service

POSSIBLE RESPONSE:

The maximum **STREAM PENETRATION ANGLE** in the fire service is considered to be **45 DEGREES**. (if the angle increases, the penetration decreases).

QUESTION:

In regards to fire streams, what will **EXCESSIVE PRESSURE** do to a small stream as compared to a large stream?

POSSIBLE RESPONSE:

EXCESSIVE PRESSURE will break-up a small stream faster than a large stream.

QUESTION:

Describe an **EFFECTIVE FIRE STREAM.**

POSSIBLE RESPONSE:

An **EFFECTIVE FIRE STREAM** discharges 90% of its volume inside a circle of 15" in diameter, and 75% of its volume inside a circle of 10" in diameter.

QUESTION:

What effect will wind be to a **FIRE STREAM**?

POSSIBLE RESPONSE:

Any wind will **HINDER** a **FIRE STREAM**.

QUESTION:

What is the **RULE OF THUMB** for the correct size of tip on different hose lay lengths of 2 1/2" hose?

POSSIBLE RESPONSE:

The **RULE OF THUMB** is:

1. Short lay = 0' to 300', use: 1 1/4" tip = 1/2 the diameter of the hose.
2. Medium lay = 300' to 600', use: 1 1/8" tip = 1st size smaller than 1/2 the diameter of the hose.
3. Long lay = 600' to 900', use: 1" tip = 2nd size smaller than 1/2 the diameter of the hose.

QUESTION:

In the fire service, what is considered to be the highest floor to direct **FIRE STREAMS** from **GROUND LEVEL?**

POSSIBLE RESPONSE:

The highest floor to direct a **FIRE STREAM** from **GROUND LEVEL** is to the **THIRD FLOOR.**

QUESTION:

What are the some of the main causes of a **DEFECTIVE** hose stream?

POSSIBLE RESPONSE:

Some of the main causes of a **DEFECTIVE** hose stream are:

1. Too little pressure.
2. Too much pressure.
3. A defective tip.
4. Air in the hose line.
5. Hose kinked or twisted near the nozzle.
6. Kinks in the hose line.

QUESTION:

As far as fire streams are concerned, at what angle does the greatest **HORIZONTAL REACH** occur?

POSSIBLE RESPONSE:

The greatest **HORIZONTAL REACH** occurs a **30 DEGREE TO 34 DEGREE** angle.

QUESTION:

As far as fire streams are concerned, at which angle does the maximum effective **VERTICAL REACH** occur?

POSSIBLE RESPONSE:

The maximum effective **VERTICAL REACH** occurs at a **60 DEGREE TO 75 DEGREE** angle.

QUESTION:

In the fire service what is considered to be the maximum **PENETRATION** distance from a hand line into a structure?

POSSIBLE RESPONSE:

The maximum **PENETRATION** distance from a hand line into a structure is considered to be **50 FEET**.

QUESTION:

In the fire service what is considered to be the smallest tip to use for **MASTER STREAMS**?

POSSIBLE RESPONSE:

Smallest tip for **MASTER STREAMS** = 1 1/4"

QUESTION:

What are some of the things that will cause a **DEFECTIVE HOSE STREAM**?

POSSIBLE RESPONSE:

Some of the things that will cause a **DEFECTIVE HOSE STREAM** are:

1. Too much pressure.
2. Too little pressure.
3. Air in hose lines.
4. Kinks in the hose.
5. Hose twisted near nozzle.
6. Defective nozzle.

QUESTION:

Why is **WATER** considered the most important extinguishing agent?

POSSIBLE RESPONSE:

WATER is the most important extinguishing agent because of its physical characteristics, its universal availability, and its low cost.

QUESTION:

What are some of the things that will affect the **COURSE** of a **FIRE STREAM**?

POSSIBLE RESPONSE:

The **COURSE** of a **FIRE STREAM** is affected by:
Gravity, Friction due to air pressure, Wind velocity, and Obstacles.

QUESTION:

What are some of the advantages and disadvantages of **SPRAY FIRE STREAMS** in contrast to solid streams?

POSSIBLE RESPONSE:

SPRAY FIRE STREAMS in contrast to solid streams:

ADVANTAGES:
1. Absorbs more heat, more rapidly.
2. Covers a greater area with water.
3. Uses less water.

DISADVANTAGES:
1. Requires higher discharge pressures.
2. Has a shorter reach.
3. Has less penetration.
4. Has less cooling effect in subsurface areas such as chard wood.

QUESTION:

What are some of the things **WATER** does as a **COOLING AGENT** for **FLAMMABLE LIQUIDS**?

POSSIBLE RESPONSE:

Some of the things that **WATER** does as a **COOLING AGENT** for **FLAMMABLE LIQUIDS**:
1. Cuts off release of vapor from surface of a high flash point oil, thus extinguishes fire.
2. Protects firefighters from flame and radiant heat.
3. Protects flame-exposed surfaces.

QUESTION:

What are some of the things that **WATER** does as a **MECHANICAL TOOL** for **FLAMMABLE LIQUIDS**?

POSSIBLE RESPONSE:

Some of the things that **WATER** does as a **MECHANICAL TOOL** for **FLAMMABLE LIQUIDS** are:
1. Controls leaks.
2. Directs the flow of the product to prevent its ignition, or to move the fire to an area where it will do less damage.

QUESTION:

What are some of the things that **WATER** does as a **DISPLACING MEDIUM** for **FLAMMABLE LIQUIDS**?

POSSIBLE RESPONSE:

Some of the things that **WATER** does as a **DISPLACING MEDIUM** are:

1. **WATER** will float oil above a leak in a tank.
2. **WATER** will cut-off the fuels escape route by pumping it into a leaking pipe ahead of the leak.

QUESTION:

Under what circumstances may **WATER** be used to smother **FLAMMABLE** or **COMBUSTIBLE** liquid fires?

POSSIBLE RESPONSE:

WATER may be used when the liquid has:

1. Flash point above 100 degrees F.
2. Specific Gravity of 1.1 or heavier.
3. When liquid is not water soluble.

QUESTION:

What determines the **EFFECTIVENESS** of any fire stream?

POSSIBLE RESPONSE:

The **EFFECTIVENESS** of any fire stream is determined by the **MOBILITY** of the stream.

QUESTION:

In the fire service, what is considered to be the definition of a **FIRE STREAM**?

POSSIBLE RESPONSE:

FIRE STREAM is a stream of water from the time it leaves a nozzle until it reaches the point of intended use.

QUESTION:

What are some of the advantages of adding a **WETTING AGENT** to water?

POSSIBLE RESPONSE:

Some of the advantages of adding a **WETTING AGENT** to water are:

1. Increases waters heat absorption ability.
2. Reduces the surface tension of the water.
3. Increases the waters penetration ability.

QUESTION:

In relation to ladders, what does **TRUSS** mean?

POSSIBLE RESPONSE:

TRUSS means to support, strengthen or stiffen as a beam.

QUESTION:

What should you do if your apparatus must be **DRIVEN** over fire hose?

POSSIBLE RESPONSE:

If the apparatus that I am responding in must be **DRIVEN** over fire hose, it is best to use a hose bridge. If it is necessary to **DRIVE** directly over fire hose without the use of a hose bridge during firefighting operations, one rule that always applies under all conditions is that the apparatus wheels should coast slowly over the hoseline. The hose should be charged with water under pressure. The most damage occurs when the fire hose is empty.

QUESTION:

What is the recommended **SEQUENCE** of operation of aerial ladders and aerial platforms?

POSSIBLE RESPONSE:

SEQUENCE of operation of aerial ladders and aerial platforms: Raise, Rotate, Extend, and Lower.

QUESTION:

Which **EXTENSION** ladders are recommend by NFPA, to be carried on a ladder truck?

POSSIBLE RESPONSE:

EXTENSION ladders recommended by NFPA, for ladder trucks:

1. One 14 foot.
2. One 28 foot.
3. One 35 foot.
4. One 40 foot.

QUESTION:

Which **STRAIGHT** ladders are recommended by NFPA, to be carried on a ladder truck?

POSSIBLE RESPONSE:

STRAIGHT ladders recommended by NFPA to be carried on a ladder truck:

1. One 16 foot.
2. One 20 foot.

QUESTION:

What percentage of time is saved by rotating and extending aerial ladders **SIMULTANEOUSLY?**

POSSIBLE RESPONSE:

NO TIME is saved by rotating and extending aerial ladders **SIMULTANEOUSLY**.

QUESTION:

What are some of the advantages of a **TRACTOR TRAILER** type aerial ladder?

POSSIBLE RESPONSE:

TRACTOR TRAILER type aerial ladder is/has more: Maneuverable, Stable, Positioned faster, Space for equipment.

QUESTION:

What is considered to be the most **STABLE** position for set-up of a tractor trailer type aerial ladder truck?

POSSIBLE RESPONSE:

The **JACKKNIFE** position of 60 degrees away from the incline position is twice as **STABLE** and considered the most **STABLE** position for aerial ladders.

QUESTION:

What is the maximum **DISTANCE** for operation from the working building that an aerial ladder should be?

POSSIBLE RESPONSE:

Aerial ladders maximum **DISTANCE** is **35 FEET** from the center of the turntable to the wall of the working building.

QUESTION:

What is considered the best **CLIMBING ANGLE** for aerial ladders?

POSSIBLE RESPONSE:

Aerial ladders best **CLIMBING ANGLE** is considered to be 70 - 80 degrees.

QUESTION:

In comparison to an aerial ladder, what are some of the **DISADVANTAGES** of an aerial platform?

POSSIBLE RESPONSE:

DISADVANTAGES of aerial platform, compared to aerial ladder:
1. Limitations of carrying capacity.
2. Limited horizontal reach.
3. Greater maintenance problems.

QUESTION:

Should an aerial ladder be **SPOTTED UPHILL**, or **DOWNHILL** from area of operation and why?

POSSIBLE RESPONSE:

Aerial ladder should be **SPOTTED DOWNHILL** from area of operation for two reasons:
1. Rung tilt will be reduced.
2. Apparatus will be more stable.

QUESTION:

When should **FIRE HYDRANTS** be inspected?

POSSIBLE RESPONSE:

FIRE HYDRANTS should be inspected **SEMIANNUALLY** and after use. Records of all test should be kept by the fire department.

QUESTION:

When should the fire department question the **CAPABILITY** of water systems?

POSSIBLE RESPONSE:

Fire departments should question the **CAPABILITY** of water systems whenever extensive construction is taking place in areas where there are mains known to be small or dead end.

QUESTION:

How are water supply systems **ANALYZED**?

POSSIBLE RESPONSE:

Water systems are **ANALYZED** by conducting **FIRE FLOW TEST**.

QUESTION:

What is the minimum **DESIRABLE** fire flow pressures from a hydrant?

POSSIBLE RESPONSE:

Minimum **DESIRABLE** fire flow pressure from a hydrant is **20 PSI**. For large/well distributed main = 10 PSI

QUESTION:

When approaching a fire hydrant, how can you visually tell the **CAPACITY RATING** of the hydrant?

POSSIBLE RESPONSE:

You can visually tell the **CAPACITY RATING** of a hydrant by the color designation:

1. Class A = green color = 1000+ GPM.
2. Class B = orange color = 500-1000 GPM.
3. Class C = red color = less than 500 GPM.

QUESTION:

What is the advantage of a **LOOPED** secondary water feeder system as compared to a non-looped water supply network?

POSSIBLE RESPONSE:

Compared to a non-looped water supply network, a secondary feeder system that is arranged in **LOOPS** increases both water supply and reliability.

QUESTION:

What is an **ADEQUATE** supply of water, so as to satisfy the fire service, demands mainly dependent upon?

POSSIBLE RESPONSE:

An **ADEQUATE** supply of water so as to satisfy the fire service demands is mainly dependent upon the arrangement and the size of the water mains.

QUESTION:

What is the PSI range of a **HIGH PRESSURE** hydrant?

POSSIBLE RESPONSE:

The PSI range of a HIGH PRESSURE hydrant is 150 PSI to 300 PSI.

QUESTION:

What is the normal **FLOW PRESSURES** of a standard fire hydrant?

POSSIBLE RESPONSE:

Normal hydrant **FLOW PRESSURE** is 65 PSI to 75 PSI.

QUESTION:

What is **RESIDUAL PRESSURE**?

POSSIBLE RESPONSE:

RESIDUAL PRESSURE is the pressure while water is flowing. Pressure on the inlet side of a pumper, or in a water main, while water is being discharged.

QUESTION:

What is **STATIC PRESSURE**?

POSSIBLE RESPONSE:

STATIC PRESSURE is the pressure when water is not flowing, therefore no pressure loss due to friction. The pressure available at discharge when no water is flowing at a hydrant, pumper, or any specific location.

QUESTION:

How can you tell if water is **FLOWING** through a sprinkler system?

POSSIBLE RESPONSE:

You can tell if water is **FLOWING** through a sprinkler system by:

1. Water flow alarm will be ringing, caused by the **FLOWING** water activating the system.
2. Water will be visible on the floors of the building and coming out of doorways, etc.

QUESTION:

What are some of the ways to **SUPPLY** a sprinkler system with water?

POSSIBLE RESPONSE:

Water **SUPPLIES** for sprinkler systems:

1. Public water works system.
2. Public and private supplies.
3. Gravity tanks. (minimum of 5000 gallons)
4. Pressure tanks. (minimum of 4500 gallons)
5. Fire pumpers and fire department connections.

QUESTION:

What are the different types of **STANDPIPE SYSTEMS**?

POSSIBLE RESPONSE:

Types of **STANDPIPE SYSTEMS**:

1. Wet : supply valve is open with water pressure at all times.
2. Dry : no permanent water supply in sprinklers.
3. Automatic supply system : opening hose valve.
4. Manual to remote system : is at hose station.

QUESTION:

Name two types of **INDICATING VALVES**:

POSSIBLE RESPONSE:

Two types of **INDICATING VALVES** are:
1. OS&Y : Outside stem and yoke.
2. PIV : Post indicator valve.

QUESTION:

Describe a **WYE CONNECTION**, and it's use.

POSSIBLE RESPONSE:

The **WYE CONNECTION** has a female stem and two male branches. The **WYE CONNECTION** is used to divide one hose line into two or more lines.

QUESTION:

Describe a **SIAMESE FITTING**, and it's use.

POSSIBLE RESPONSE:

The **SIAMESE FITTING** has a male stem and two female branches. The **SIAMESE FITTING** is employed to bring two or more hose lines into a single line.

QUESTION:

In relation to the fire, where should the **SIAMESE FITTING** be located, when in use?

POSSIBLE RESPONSE:

The **SIAMESE FITTING**, when used should be as close to the fire as possible.

SECTION 4

JOB KNOWLEDGE

RESPONSIBILITIES

QUESTION:

What are the **RESPONSIBILITIES** and **DUTIES** of a **FIRE CAPTAIN**?

POSSIBLE RESPONSE:

The **RESPONSIBILITIES** and **DUTIES** of a **FIRE CAPTAIN:**

1. Be aware of the departments policy and procedures.
2. Be aware of the departments rules and regulations.
3. Respond to emergencies:
   - Fires
   - Rescues
   - Haz Mat
   - ETC.
4. Command emergencies until relieved by a superior officer.
5. Supervise subordinate Firefighters and other personnel in station and in the field.
6. Train Firefighters and other personnel.
7. Make ready apparatus, equipment and personnel.
8. Fire investigation.
9. Records, reports and related paperwork.
10. Public relations.
11. The implementation of new procedures.
12. Administration task.
13. Tours of department and equipment.
14. To maintain a through knowledge of Fire Department Public relation programs.
15. To stay current of:
    - Water supply and hydrants.
    - Firefighting techniques.
    - Fire Prevention techniques.
    - First aid techniques.
16. To be involved in :
    - Fire prevention.
    - Pre-fire plans.
    - Hydrant maintenance.
17. Related duties as assigned.

ADD ANYTHING THAT'S APPROPRIATE FOR YOUR DEPARTMENT
(read job announcement flyer)

QUESTION:

What do you feel is the most important **RESPONSIBILITY** of a **FIRE CAPTAIN**?

POSSIBLE RESPONSE:

SEE BELOW & RESPOND WITH YOUR OWN THOUGHTS

QUESTION:

What are the **RESPONSIBILITIES** of a **FIRE CAPTAIN** at a fire?

POSSIBLE RESPONSE:

The **RESPONSIBILITIES** of a **FIRE CAPTAIN** at a fire:

1. Safety of personal.
2. roper use of apparatus and equipment.
3. Command fire until relieved by a superior officer.
4. Merging of assignments with other companies to minimize efforts towards the same goals.
5. Be aware of:
   Amount of water being used and available.
   Other pumpers and apparatus.
   Hydrant locations, water sources.
   Other apparatus on scene or responding.
   Equipment available .
   Radio communications.
   Location of hose lines: ie: Inside or Outside, Upper elevations, Ground floors, basements, ETC.

ADD WHA IS APPROPRIATE FOR YOUR DEPARTMENT

QUESTION:

What are some of the most important facets of the position of **FIRE CAPTAIN** on your department?

POSSIBLE RESPONSE:

Some of the most important facets of the position of **FIRE CAPTAIN** are:

1. The preservation of my own and co-workers lives.
2. The preservation of apparatus and equipment.
3. Training.
4. Pre-fire planning, Fire Prevention and Suppression.

ADD YOUR OWN THOUGHTS

SUPERVISION

QUESTION:

Can you briefly define **SUPERVISION** in the fire service?

POSSIBLE RESPONSE:

SUPERVISION is the **WATCHING-OVER** process.

QUESTION:

What type of **LEADERSHIP SKILLS** do you feel are the most important for the position of **FIRE CAPTAIN**, and would you be more democratic or authoritarian in your leadership?

POSSIBLE RESPONSE:

I think that the ability to understand and get along with subordinates is one of the most important leadership skill for a Fire Captain to poses along with good training skills.

QUESTION:

How would you handle your close friends if you were promoted to **FIRE CAPTAIN** and stayed on the same shift?

POSSIBLE RESPONSE:

RESPOND WITH YOUR OWN IDEAS.

QUESTION:

As **FIRE CAPTAIN**, what would you expect of your men and how would you communicate this to them?

POSSIBLE RESPONSE:

As **FIRE CAPTAIN** I would communicate face to face to my subordinates that I expect them to perform the duties of their position in an acceptable manner.

ADD YOUR OWN IDEAS

QUESTION:

What methods will you use to handle **PROBLEM EMPLOYEES**?

POSSIBLE RESPONSE:

In a situation where I would have to handle a problem employee, I would **COMMUNICATE** face to face with the employee and try to find out why he is not performing properly. I would try to motivate this employee so that he might enjoy his job more.

ADD YOUR OWN IDEAS

QUESTION:

Tell us what you know about the **SKELLEY DECISION** and what role does it play in your job as a **FIRE CAPTAIN**?

POSSIBLE RESPONSE:

The **SKELLEY DECISION** declares that Firefighters of permanent positions:

1. Has a property interest in his/her employment.
2. Is entitled to **DUE PROCESS** to protect his/her interest.
3. Cannot be disciplined without including pre-removal safeguards, including:
   Notice of proposed action. (written)
   Reason for action.
   Copy of allegation.
   Facts on which allegation is based upon.
   The right to reply to the authority originating the discipline. (written or verbally)

The **SKELLEY DECISION** pertains to the state of California, but will more than likely become the standard for the remaining states.

QUESTION:

Describe **AUTOCRATIC LEADERSHIP** in one word. (also **DEMOCRATIC** and **LAISSEZ-FAIR**)

POSSIBLE RESPONSE:

AUTOCRATIC = **BOSS**
DEMOCRATIC = **PARTICIPATIVE**
LAISSEZ-FAIR = **FREE-REIGN**

QUESTION:

How will you **MOTIVATE** your subordinates?

POSSIBLE RESPONSE:

Subordinates can be motivated by regular face to face communication and assignment of task that they have some expertise in. try to make subordinates feel comfortable in performing their task.

QUESTION:

What methods will you use to increase good **MORALE**?

POSSIBLE RESPONSE:

I will increase **MORALE** with face to face communication on a regular basis. I will involve personnel in task that they have some expertise in and feel comfortable performing. High **MORALE** can be attained by a supervisor being fair and impartial!

ADD YOUR OWN THOUGHTS

QUESTION:

Can you tell us one way that a you can tell that a **FIRE CAPTAIN** is doing a good job as a **SUPERVISOR**?

POSSIBLE RESPONSE:

One way that you can tell that a **FIRE CAPTAIN** is doing a good job as a **SUPERVISOR** is by **GOOD MORALE**.

QUESTION:

As a **FIRE CAPTAIN**, what might laziness, griping, discontent, and tardiness from your men indicate to you?

POSSIBLE RESPONSE:

Laziness, griping, discontent, tardiness are all indications of **LOW MORALE**.

QUESTION:

As a **FIRE CAPTAIN**, what would be your definition of **MORALE?**

POSSIBLE RESPONSE:

MORALE = "ESPRIT DE CORE". A state of mind, good spirits with positive hope.

QUESTION:

As a company officer, what will be your **MANAGEMENT STYLE?**

POSSIBLE RESPONSE:

My **MANAGEMENT** style will be a blend of **AUTOCRATIC** and **DEMOCRATIC**. During emergency situations it will normally be **AUTOCRATIC**. I will represent management to my subordinates and at the same time serve both supervisors and subordinates. A coordinator between the men and the management.

INSERT YOUR OWN MANAGEMENT STYLE

QUESTION:

In your opinion are **FIRE CAPTAINS** part of the **MANAGEMENT TEAM?**

POSSIBLE RESPONSE:

Fire Captains are company supervisors that represent management to his subordinates, while serving both.

QUESTION:

If you as a **FIRE CAPTAIN** criticize a subordinate, what should accompany this criticism?

POSSIBLE RESPONSE:

A possible remedy should always accompany criticism.

QUESTION:

Describe the City of __________ **ADMINISTRATIVE POLICY** governing **DISCIPLINARY ACTION**, specifically, the responsibility of a **FIRE CAPTAIN**, and the parameters of his authority.

POSSIBLE RESPONSE:

DESCRIBE YOUR CITIES DISCIPLINARY ACTION POLICY

QUESTION:

What is the relationship between **SUPPRESSION** and **FIRE SAFETY CONTROL** on the __________ Fire Department?

POSSIBLE RESPONSE:

The primary task concerning **FIRE SAFETY CONTROL** on the __________ Fire Department is security against fire and its effects. "SAFETY FIRST AND ALWAYS"
INSERT YOUR DEPARTMENTS POLICY

QUESTION:

Briefly, what should **DISCIPLINE** do to a firefighter?

POSSIBLE RESPONSE:

DISCIPLINE should change the firefighters attitude and habits toward his job performance.

QUESTION:

What is the difference between a **MANAGER** and **SUPERVISOR**?

POSSIBLE RESPONSE:

A **MANAGER** is the individual that gets things done through others. A **SUPERVISOR** is the individual that represents management to his subordinates while serving both. The **SUPERVISOR** guides and controls his subordinates as a coordinator between them and management. **INSERT YOUR OWN DEFINITION**

QUESTION:

The **PERSONAL PROBLEMS** of a subordinate would be the concern of you as a **FIRE CAPTAIN** under what circumstances?

POSSIBLE RESPONSE:

The **PERSONAL PROBLEMS** of subordinates are usually of proper concern to a **FIRE CAPTAIN** only in so far as these problems affect the job performance of the subordinate.

QUESTION:

What is **DISCIPLINE** in the Fire Service?

POSSIBLE RESPONSE:

In the Fire Service **DISCIPLINE** could be described as **EDUCATION IN CONDUCT**.

DISCIPLINE is the standard of conduct that is used to direct Fire Service personnel in the performance of their duties.

DISCIPLINE should not be thought of as punishment.

QUESTION:

A **FIRE CAPTAIN** as the supervisor can usually keep trouble to a minimum by doing what?

POSSIBLE RESPONSE:

A **FIRE CAPTAIN** can usually keep trouble down by foreseeing trouble and making changes before it arises.

QUESTION:

As **FIRE CAPTAIN**, what method could you use to get along better with your men?

POSSIBLE RESPONSE:

One method that I could use to get along better with the men would be to make sure that my disposition fits the atmosphere of the moment.

QUESTION:

As **FIRE CAPTAIN**, what is usually considered the best way to handle a change in work procedure?

POSSIBLE RESPONSE:

As **FIRE CAPTAIN**, I would discuss the work procedures with the men and listen to any ideas that they may have.

QUESTION:

In your opinion, what is the main value of **EFFICIENCY RATINGS**? (performance evaluations)

POSSIBLE RESPONSE:

In my opinion, the main value of **EFFICIENCY RATINGS** is that they let the men know where they stand and why.

QUESTION:

What is the most significant **DIFFERENCE** between operating and service functions of an administrative organization?

POSSIBLE RESPONSE:

The most significant difference is that operating activities are an end in themselves while service activities are a means to an end.

QUESTION:

What is the first step that a Fire Department can take in order to **IMPROVE** upon its present **PUBLIC RELATIONS** program?

POSSIBLE RESPONSE:

The Fire Department can improve upon its present public relations program by first finding out what the **VIEWS** of the public are towards the Department.

QUESTION:

What factors, within a community, should a Fire Officer be aware of that will influence the actions of community groups?

POSSIBLE RESPONSE:

The factors, within a community, that a Fire Officer should be aware of are:

- Economics.
- Religious.
- Race-ethnic.
- Age.
- Organizations.
- Family structures.

QUESTION:

Where does the responsibility for public attitudes form within the Fire Service?

POSSIBLE RESPONSE:

All Fire Officers and Firefighters are responsible for the formation of beneficial public attitudes towards the Fire Department.

QUESTION:

What's the goal of the Fire Service to a community?

POSSIBLE RESPONSE:

The goal of the Fire Service, to the community, is **SERVICE**.

TRAINING

QUESTION:

As a **FIRE CAPTAIN**, what would be your first task in establishing an effective **TRAINING PROGRAM**?

POSSIBLE RESPONSE:

In order to establish an effective **TRAINING PROGRAM**, my first task would be to determine the objectives.

QUESTION:

As the **FIRE CAPTAIN**, what is the first thing that you would **TEACH** a new recruit?

POSSIBLE RESPONSE:

As **FIRE CAPTAIN**, the first thing that I would **TEACH** a new recruit is the concept of team work, no room for free lance operations in the fire service, we must operate as a team to accomplish the task given, provide for maximum safety for all and to be successful in all areas of a Firefighter's responsibilities.

QUESTION:

What are some of the indications that a Firefighter needs **TRAINING**?

POSSIBLE RESPONSE:

Some of the indications that a Firefighter needs **TRAINING** are:

1. Excessive absences.
2. Unusual tardiness.
3. Lack of cooperation.
4. Poor morale

QUESTION:

As the **FIRE CAPTAIN**, what should you remember about the principle that applies to the **LEARNING** Firefighters work routines?

POSSIBLE RESPONSE:

One principle that always applies is that the **LEARNING** rate is different in each person.

QUESTION:

In your opinion, what is the disadvantage of a Firefighter **TRAINING** himself?

POSSIBLE RESPONSE:

Probably the biggest disadvantage of a Firefighter **TRAINING** himself would be the fact that he most likely will not learn the correct way.

QUESTION:

In your opinion what is the most important asset that a **TRAINING OFFICER** should have?

POSSIBLE RESPONSE:

TRAINING OFFICER'S must have a knowledge and understanding of the job.

QUESTION:

Assume that you have been assigned to **TRAIN** a group on a new apparatus. What would the criteria be for primary consideration when planning your course of instruction?

POSSIBLE RESPONSE:

The criteria that I would use for **TRAINING** a group on a new apparatus mainly would be: what should the group learn in this **TRAINING** session.

QUESTION:

Assume that you as a **FIRE CAPTAIN** are giving instruction on a new tool, you are asked by a member why the operation of the tool was not performed in a different manner. The method suggested by the member includes some actions which could cause injury. How would you answer the member?

POSSIBLE RESPONSE:

As the **FIRE CAPTAIN** it would be my duty to point out incorrect methods of use, so that they may be avoided. I would inform the member as to why this different manner of use is unsafe.

QUESTION:

As a **FIRE CAPTAIN** you observe a man performing a task improperly. When you discuss this with him, you are told that his previous **CAPTAIN** taught him this method. What would be the best action to take at this point?

POSSIBLE RESPONSE:

As **FIRE CAPTAIN** it would be best for me to indicate to the subordinate as to why the method he is using is not acceptable.

QUESTION:

In your opinion how should a new evolution be **TAUGHT**?

POSSIBLE RESPONSE:

I think that new evolutions should be **TAUGHT** slowly, and repeated as often as necessary.

QUESTION:

In your opinion, how often should **TRAINING** techniques be reviewed and modified?

POSSIBLE RESPONSE:

TRAINING techniques should be reviewed continuously and modified as necessary.

QUESTION:

Assume that a suggestion is made to you, the **FIRE CAPTAIN**, from one of your Firefighters for improving a firefighting procedure. In your opinion what would be the best action to take regarding this suggestion?

POSSIBLE RESPONSE:

In my opinion as the **FIRE CAPTAIN**, if it is a good suggestion, I should pass it on to my superior.

QUESTION:

As a **FIRE CAPTAIN** you should be aware of different methods of instruction for your subordinates. What would be your opinion of another **FIRE CAPTAINS** advice to a new firefighter, who has a question about a piece of equipment, telling him to watch the older men at a fire and to ask them later if he had any doubts?

POSSIBLE RESPONSE:

This would be **POOR** advice, because the **TRAINING** of members is primarily a responsibility of the **FIRE CAPTAIN**.

QUESTION:

Assume that you as a **FIRE CAPTAIN** are **INSTRUCTING** a Firefighter in a new procedure. The Firefighter appears to be puzzled but does not ask any questions. In your opinion what would be the best way to handle this situation?

POSSIBLE RESPONSE:

In my opinion the best way to handle this situation would be to ask the Firefighter to explain the procedure in his own words.

QUESTION:

What would be your course of action if you as **FIRE CAPTAIN** have a new Firefighter that appears to lack confidence in the performance of his duties?

POSSIBLE RESPONSE:

As the **FIRE CAPTAIN**, I would give the Firefighter an assignment which I thought he would be able to perform well.

QUESTION:

Assume you as a **FIRE CAPTAIN** overhear another **FIRE CAPTAIN** explaining a certain evolution to a Firefighter and you know that the procedure is about to be changed. In your opinion what should you do?

POSSIBLE RESPONSE:

In my opinion it would be best to **IGNORE** the conversation and wait until everyone is formally notified of the new evolution.

QUESTION:

As the **FIRE CAPTAIN**, what should you do in order to make it easier for a new Firefighter to remember how a specific piece of work is to be done?

POSSIBLE RESPONSE:

As the **FIRE CAPTAIN**, I should explain the reasons for doing the work in the manner specified.

QUESTION:

What would you do if as **FIRE CAPTAIN** you observed one of your Firefighters commit a grave error at a fire?

POSSIBLE RESPONSE:

As a **FIRE CAPTAIN**, it would be my responsibility to instruct the Firefighter so that it does not happen again.

QUESTION:

In your opinion, is it good practice to keep a close check on a new evolution the first few times that it is used? Why?

POSSIBLE RESPONSE:

Yes, it is a good practice, because it will give an opportunity to determine if the evolution will require any modifications.

QUESTION:

Assume that you are the **FIRE CAPTAIN** giving **TRAINING** to members on proper courtroom conduct and behavior in a jury trial. One of your crew would like to know if it is proper to use technical jargon when testifying. What would be your advice to this Firefighter?

POSSIBLE RESPONSE:

I would suggest to the Firefighter that it is alright to use technical jargon, but that he should follow up the technical jargon by explaining it in simple terms.

QUESTION:

As **FIRE CAPTAIN**, what would be your criteria for selecting a member from your company to be trained to operate a new type of complicated mechanical device.

POSSIBLE RESPONSE:

As a general rule, I think the best criterion would be to select a member that has acquired the greatest skill in the operation of other mechanical devices.

QUESTION:

Name the two techniques of **PROGRAMMED LEARNING**.

POSSIBLE RESPONSE:

The two techniques of **PROGRAMMED LEARNING** are **LINEAR** and **BRANCHING**.

QUESTION:

As the **FIRE CAPTAIN**, what kind of reaction would you expect if in reviewing a fire with your crew you presented it in a way that you cited it as an example of very poor work on their part?

POSSIBLE RESPONSE:

This would be an **UNDESIRABLE** approach to use by any **FIRE CAPTAIN**, the response of the crew is likely to be a strong defense of their work.

QUESTION:

If you as **FIRE CAPTAIN** were conducting **TRAINING** with your crew and one of the members is asking what appears to be an unnecessary number of questions, what is the most probable explanation for this?

POSSIBLE RESPONSE:

In this situation, the most probable explanation is that I, as the **FIRE CAPTAIN** and instructor am not making the instructions clear.

QUESTION:

As **FIRE CAPTAIN**, what is are your responsibilities, in regards to management, pertaining to training your crew?

POSSIBLE RESPONSE:

In this situation as Fire Captain my responsibility for training my crew is advantageous for management because it will facilitate placing the responsibility for the crews performance with me as the Fire Captain.

QUESTION:

An older Firefighter, who has been consistently passed over for promotion to **FIRE CAPTAIN**, has given up studying and become lazy in his work assignments at the firehouse. How would you handle this situation?

POSSIBLE RESPONSE:

In this situation I would discuss the matter with the Firefighter and try to re-stimulate him by finding out his interest and expertise.

QUESTION:

A new Firefighter is having trouble studying and performing duties required and is too shy to ask for help. How will you as **FIRE CAPTAIN** get this Firefighter to study and perform properly?

POSSIBLE RESPONSE:

In this situation I would discuss face to face with the individual methods of study and performance of duties whenever giving assignments.

QUESTION:

Assume that you as the **FIRE CAPTAIN** were holding a **DRILL** on the fire codes and want your crew to remember a particularly significant idea. What method would you use to help your crew remember?

POSSIBLE RESPONSE:

I would **REPEAT** the idea four or five times during the course of the drill.

QUESTION:

What methods would you use in the **TRAINING** of new Firefighters?

POSSIBLE RESPONSE:

In the training of new Firefighters I would use "HANDS ON" method of teaching as much as possible.

ADD TOUR OWN IDEAS

QUESTION:

What is **REQUIRED FIRE FLOW?**

POSSIBLE RESPONSE:

REQUIRED FIRE FLOW is the amount of water needed for fire fighting purposes in order to confine a major fire to the buildings within a block or other group complex.

QUESTION:

What is the **MINIMUM FIRE FLOW?**

POSSIBLE RESPONSE:

MINIMUM FIRE FLOW is 500 GPM for un-congested areas of small buildings.

QUESTION:

What is the **MAXIMUM REQUIRED FIRE FLOW?**

POSSIBLE RESPONSE:

MAXIMUM REQUIRED FIRE FLOW is 12,000 GPM for large areas (industrial) and downtown areas of large cities.

An additional 2,000 GPM to 8,000 GPM are required for simultaneous fires.

Total **MAXIMUM REQUIRED FIRE FLOW** is **20,000 GPM**.

QUESTION:

What is the most common **SPREAD OF FIRE** in buildings?

POSSIBLE RESPONSE:

The most common **SPREAD OF FIRE** in buildings is unprotected vertical openings.

QUESTION:

What are some of the characteristics of a **BACKDRAFT** situation?

POSSIBLE RESPONSE:

Some **BACKDRAFT** characteristics are:
1. Smoke under pressure.
2. Dense grayish, yellowish smoke.
3. Puffing smoke from cracks, moving up rapidly.
4. Confinement of excessive heat.
5. Sweating windows, hot to the touch and dark in color.
6. Muffled sounds.
7. No visibly flame.
8. Rapid movement of air inward when an opening is made.

QUESTION:

What are the **EIGHT** basic divisions of **FIRE STRATEGY**?

POSSIBLE RESPONSE:

The **EIGHT** divisions of **FIRE STRATEGY** are:
1. Size-up.
2. Rescue.
3. Extinguishment.
4. Ventilation.
5. Salvage.
6. Confinement.
7. Exposure.
8. Overhaul.

QUESTION:

What are the two divisions of **FIRE STRATEGY** that can be initiated whenever they are needed?

POSSIBLE RESPONSE:

The two divisions of **FIRE STRATEGY** that may be initiated whenever they are needed are:
1. Ventilation. 2. Salvage.

QUESTION:

One of the most important tools an Incident Commander (I/C) can have is a good understanding of **STRATEGY** and **TACTICS**. Can you tell us the differences between the two and explain their roles at an incident?

POSSIBLE RESPONSE:

STRATEGY: involves the development of a basic plan to most effectively deal with a situation. The plan must identify major goals and prioritize objectives for the tactical elements. **STRATEGIC** decisions are based upon an evaluation of the situation, the risk potential, and the capabilities of available resources. Identify priorities (ie. rescue, fire control, property conservation). The **STRATEGIC** plan identifies where and when the forces will attempt to control the fire and how their activities will be combined and prioritized or defensive attack plan.

TACTICS: are the methods selected by the fireground Commander to implement the **STRATEGIC** plan. The **TACTICAL** objectives define specific functions that are assigned to groups or companies operating under company or sector officers. The achievement of these objectives contributes to the **STRATEGIC** goals and must be compatible with the overall **STRATEGIC** plan. Incident **TACTICS** usually involve a coordinated mixture of task directed toward the **STRATEGIC** goals. **TACTICAL** functions include:

1. Search and rescue.
2. Exposure protection.
3. Confinement.
4. Extinguishment.
5. Ventilation.
6. Property conservation & Overhaul.

QUESTION:

What does the term **SIZE-UP** refer to?

POSSIBLE RESPONSE:

SIZE-UP is the mental evaluation made by the fire officer in charge, which enables him to determine a course of action. It includes such factors as time, location, nature of occurrence, life hazard, exposure, property involved, nature and extent of the fire, available water supply, and other fire fighting facilities. **SIZE-UP** is a report usually via radio, giving existing conditions of an emergency. **SIZE-UP** is continuous throughout the entire fire fighting operation.

QUESTION:

What are the four major forms of **STRATEGY** and how are they utilized?

POSSIBLE RESPONSE:

The four major forms of **STRATEGY** are:
1. OFFENSIVE; utilized when the fire is small or when an attack is made directly on the seat of the fire. (aggressive, close attack followed-up with other forces for support)
2. DEFENSIVE; utilized by protecting exposures without the advancing of heavy hose streams.
3. OFFENSIVE-DEFENSIVE; Stop extension of fire with back-up support after an aggressive fire attack first on the major involvement of fire. (blitz attack = making a strong fire attack while setting-up defensive lines)
4. DEFENSIVE-OFFENSIVE; start by taking holding actions to protect exposures, then utilize incoming units, equipment, and manpower to initiate an aggressive fire attack.

QUESTION:

What factors in the **SIZE-UP** process would a Fire Officer have to consider before forming a plan of operation?

POSSIBLE RESPONSE:

The Fire Officer would have to consider:
1. Possibilities = facts.
2. Probabilities = resources.

QUESTION:

What are the four stages of **SIZE-UP**?

POSSIBLE RESPONSE:

The four stages of **SIZE-UP** are anticipating the situation, gathering of the facts, and the determination of the procedure.

QUESTION:

Probably, the most important task a fire ground Commander must accomplish for any incident is **SIZE-UP**.

A proper **SIZE-UP** is accomplished using a combination of knowledge and experience and becomes a determining factor in choosing the proper strategic plan and implementing the most effective tactical objectives.

Can you tell us what a **SIZE-UP** is and how it can be utilized? In your answer please include: who is responsible for **SIZE-UP,** when does a **SIZE-UP** begin, when is it completed and what are some of the elements of a **GOOD SIZE-UP**?

POSSIBLE RESPONSE:

SIZE-UP is the evaluation made by the officer in charge of an incident which enables him to accomplish his mission. It is the responsibility of the officer in charge of the first alarm units and becomes the responsibility of any officer who may later take charge of operations at a fire or other emergency.

A commanding officer must:

1. Quickly survey and analyze the situation.
2. Weigh the various factors.
3. Apply basic principles.
4. Decide what action should be taken.
5. Formulate a plan of operation.
6. Exercise command.

Success or failure on the fireground depends to a major degree upon the ability of a commanding officer to perform these functions in a practical and skillful manner. This requires a disciplined mind which has been trained in the art of clear and logical thinking.

Some of the elements of a **GOOD SIZE-UP** include:

1. FACTS
    - A. Time.
    - B. Location.
    - C. Nature of emergency.
    - D. Life hazard.
    - E. Exposures.
    - F. Building or buildings involved.
    - G. Fire.
    - H. Weather.

2. PROBABILITIES
    A. Life hazard.
    B. Fire extension.
    C. Explosions.
    D. Building collapse.
    E. Weather changes.
    F. Preventable damage.

3. OWN SITUATION
    A. Personnel and equipment.
    B. Additional assistance available.
    C. Water supply.
    D. Private fire protection.
    E. Action already taken.

4. DECISION
    A. Initial decision.
    B. Supplemental decision.

5. PLAN OF OPERATION
    A. Orders and instructions.
    B. Supervision of operations.

QUESTION:

When does **SIZE-UP** begin?

POSSIBLE RESPONSE:

SIZE-UP begins with the routine fire inspection process of pre-fire plans and fire inspections.

QUESTION:

What is the difference between a **PROBABILITY** and a **POSSIBILITY**?

POSSIBLE RESPONSE:

PROBABILITY: probably will happen and **POSSIBILITY**: what could happen.

QUESTION:

Briefly what is meant by the term **FIRE TACTICS**?

POSSIBLE RESPONSE:

FIRE TACTICS: the various maneuvers that can be used in a strategy to successfully fight a fire.

QUESTION:

What should a **PLAN OF OPERATION** consist of?

POSSIBLE RESPONSE:

A **PLAN OF OPERATION** should consist of:

1. Strategy.
2. Staff support.
3. Communication.
4. Command in charge.
5. Command position.
6. Incoming aid.

QUESTION:

What DATA is acquired at the receipt of an alarm?

POSSIBLE RESPONSE:

The **DATA** that is received at the time that an alarm comes in is:
1. Time of day.
2. Location of incident.
3. Type of incident.
4. Responding units.
5 Weather conditions.

QUESTION:

As far as **RESOURCES**, what will you as a **FIRE OFFICER** be thinking about concerning an incident?

POSSIBLE RESPONSE:

As a **FIRE OFFICER** the **RESOURCES** that I will be thinking about concerning an incident are:

1. APPARATUS:
   - A. Type and number of apparatus for initial response.
   - B. Total type and number apparatus available.
   - C. Type of hose loads from mutual aid.
   - D. Pump capabilities.
   - E. ETC.
2. PERSONNEL:
   - A. Manpower on scene.
   - B. Total manning levels.
   - C. Performance capabilities.
   - D. Organization capabilities.(staff support)
   - E. Manpower level of training.
3. EQUIPMENT:
   - A. Type of inventory.
   - B. Type of specialized equipment.(foam etc.)
   - C. Specialized skills needed for equipment.
4. COOPERATIVE AGENCIES:
   - A. Police.
   - B. Utilities. (gas, water, electric)
   - C. Public works.
   - D. Civilians.
5. AGENTS AND APPLICATION:
   - A. Availability of water, foam, powders, etc. (extinguishing agents)
6. OPERATIONS, EFFECTIVENESS OF BUILT IN DEVICES:
   - A. Sprinkler systems.
   - B. Standpipe systems.
   - C. Detectors.
   - D. Alarms, etc.
   - E. Air conditioning ducts.
   - F. Smoke and heat vents.
   - G. Fire doors.
   - H. Fire walls.

ADD ANY OF YOUR OWN APPROPRIATE IDEAS

QUESTION:

What is the **DATA** acquired upon arrival to an incident?

POSSIBLE RESPONSE:

The **DATA** acquired upon arrival to an incident is:

1. The true nature of the incident.
2. Actual access to incident.
3. Determination of actual smoke and fire conditions.
4. Nature of the threat.
5. The operational objectives that I want to accomplish.

QUESTION:

What is a **COMMAND POST**?

POSSIBLE RESPONSE:

The **COMMAND POST** is the physical location where information can be focused and transmitted so that results can be evaluated.(the I.C. is located here)

QUESTION:

Where should the **COMMAND POST** be located?

POSSIBLE RESPONSE:

The **COMMAND POST** should be located any where that a loss of perspective (tunnel vision) can be avoided. No hard fast rules as to the location. The I.C. generally should be in front of the incident. He should assign someone to the rear of the incident and remaining sector areas.

QUESTION:

What is the definition of **BASIC POSITIONS**?

POSSIBLE RESPONSE:

The **BASIC POSITIONS** are the generally accepted locations where equipment and personnel are displayed to prevent the emergency from escalating.

QUESTION:

What would you consider the basic guidelines to follow when coordinating **RESCUE** efforts?

POSSIBLE RESPONSE:

The basic guidelines to follow for **RESCUE** efforts would be to have a plan and keep the procedure orderly.

QUESTION:

What are the basic positions of an **ENGINE COMPANY** at an incident?

POSSIBLE RESPONSE:

The basic positions of an **ENGINE COMPANY** at an incident are:
1. The front of emergency.
2. The rear of emergency.
3. The interior to stop vertical spread.
4. The interior toward horizontal spread.
5. Specifically to exposures.
6. At water supply.
7. For reserve.

QUESTION:

What are the basic positions for **TRUCK COMPANIES** at an incident?

POSSIBLE RESPONSE:

The basic positions for a **TRUCK COMPANY** at an incident are:
1. Ladder front.
2. Ladder rear.
3. Perform ventilation.
4. Ladder exposures.
5. Salvage.
6. For reserve.

QUESTION:

Why is it important to closely coordinate **VENTILATION** actions?

POSSIBLE RESPONSE:

It is important for the **VENTILATION** procedures to be closely coordinated so that the overall plan of fire attack will work properly.

QUESTION:

What factors can effect the **EXTENSION OF FIRE** within a fire structure?

POSSIBLE RESPONSE:

Some of the factors that can effect the **EXTENSION OF FIRE** within a structure are:
1. The plan of fire attack.
2. Contents of the structure.
3. Horizontal openings of structure.
4. Vertical openings of structure.
5. Ventilation methods and location.

SECTION 5

SITUATION QUESTIONS

SUPERVISION

QUESTION:

In your opinion what does good **LINE SUPERVISION** imply?

POSSIBLE RESPONSE:

Good **LINE SUPERVISION** implies delegation of authority and responsibility and the sharing with subordinates of the credit for success.

QUESTION:

How would you deal with **RESENTMENT** from your peers when you are promoted and they are not?

POSSIBLE RESPONSE:

In this situation I would try to gain respect and cooperation with these men by consulting with them, face to face, on details with which the have a lot of experience.

QUESTION:

Assume that you took the Fire Captain's promotional exam with men senior to you, you obtain the highest score and are promoted. What would be the best way to work with these other men?

POSSIBLE RESPONSE:

In this situation I would deal with all of the men alike and utilize their knowledge.

QUESTION:

After being promoted to Fire Captain, how would you handle the **PROBLEM** where personnel tend to oppose and take exception to changes or innovations of the existing procedures?

POSSIBLE RESPONSE:

In this situation I would try to generate approval by supplying details pertaining to these procedures before they are instituted.

QUESTION:

Assume that you are the new Captain in a Firehouse where you are not familiar with the crew that has been working together for quite awhile. What would be your method of **CHANGEOVER** to your leadership?

POSSIBLE RESPONSE:

In this situation I would let the crew continue working at their present assignments until I am able to get to know them better.

QUESTION:

Assume that you are a **NEWLY PROMOTED** Fire Captain, what would be your method to attain your goals and gain the good will of your crew?

POSSIBLE RESPONSE:

In this situation I would continue the policies of my predecessor and gradually introduce the necessary changes to attain my goals.

QUESTION:

As far as **ATTITUDE** goes, what would be your method of securing the most efficient work from your crew?

POSSIBLE RESPONSE:

To secure the most efficient work from my crew I would strive to consistently maintain a friendly but firm attitude.

QUESTION:

Assume that you have recently been assigned as Fire Captain and you have taken an instantaneous liking to part of your crew, but one member seems to antagonize you. How would you handle this situation in respect to this member?

POSSIBLE RESPONSE:

In this situation it would be useful for me to establish from my Lieutenant his knowledge of this member and his standing as a Firefighter.

QUESTION:

If the Fire Chief were to instruct you to implement an **UNPOPULAR POLICY**, which you personally disagree with, how would you institute this policy?

POSSIBLE RESPONSE:

In this situation I would let the Fire Chief know my point of view before carrying out his instructions.

QUESTION:

As a Fire Captain, how would you develop acceptance of new **WORKING PROCEDURES** among members of your crew other than just compliance?

POSSIBLE RESPONSE:

In this situation I would call a formal or informal meeting with the crew and discuss the changes and the reasons for them.

QUESTION:

What is the term used to describe the principle for the number of Firefighters that may be supervised?

POSSIBLE RESPONSE:

The term used to describe the principle for number that may be supervised is called **SPAN OF CONTROL**.

QUESTION:

What are the 3 **DIVISIONS** of Fire Department administration?

POSSIBLE RESPONSE:

The 3 divisions are: **LINE, AUXILIARY,** and **STAFF**.

QUESTION:

Assume that you are the Fire Captain of an engine company and it is necessary for you to have a meeting with your crew to **DISCUSS A PROBLEM**. What is the best procedure to use as far as letting your subordinates express their views?

POSSIBLE RESPONSE:

In this situation it is best to let the crew express their own views first so that they will express their true feelings.

QUESTION:

As Fire Captain you have a Firefighter that comes to you about a **PERSONAL MATTER**, how would you handle this situation?

POSSIBLE RESPONSE:

In this situation I would listen courteously, and with guiding inquires encourage the Firefighter to contemplate the dilemma and come to his own solution.

QUESTION:

Assume that you are a newly appointed Fire Captain and you make a **MISTAKE** in your work, that if discovered might cause embarrassment and annoyance to your supervisors, how would you handle this situation?

POSSIBLE RESPONSE:

In this situation I would report the mistake to my immediate supervisor as soon as I was aware of the mistake.

QUESTION:

As a Fire Captain, how would you handle the distribution of assignments to a subordinate that may be undergoing severe **EMOTIONAL STRAIN**, Why?

POSSIBLE RESPONSE:

In this situation I would take into consideration of the emotional state of the crew member, because in some situations he may not be dependable.

QUESTION:

As a Fire Captain you have a Firefighter that is unhappy with a decision you have made. You consider the matter unimportant but it appears to be extremely important to the Firefighter. The Firefighter gets very excited and irritated, How would you handle this situation?

POSSIBLE RESPONSE:

In this situation I would let the Firefighter vocalize until "he gets it off his chest" and then explain the basis for my decision.

QUESTION:

What is the basic tool that a Fire Captain has to use for the **EVALUATION** of completed task?

POSSIBLE RESPONSE:

The basic tool is **ANALYSIS**.

QUESTION:

Assume that you as Fire Captain have a Firefighter that **DISAGREES** with your **EVALUATION** of his work, how would you handle this situation?

POSSIBLE RESPONSE:

In this situation I would demonstrate the basis of the evaluation and review it with the Firefighter.

QUESTION:

Assume that you are a younger, **LESS EXPERIENCED** Fire Captain at a station with another Fire Captain that constantly tells you that your ideas for improving efficiency will not work. How would you handle this situation?

POSSIBLE RESPONSE:

In this situation I would try to gain cooperation or at least an agreement to discuss these matters on an equal basis with the older Fire Captain. Also I would have to recognize that he does have a different point of view that must be taken into consideration.

QUESTION:

Assume that you as a Fire Captain have been give an assignment, by your supervisor, to be completed within a **SPECIFIC TIME FRAME**. What action would you take if after starting the assignment you realize that you are not going to meet the time frame?

POSSIBLE RESPONSE:

In this situation I would advise my supervisor of the progress of the assignment and ask for further instructions.

QUESTION:

Assume that you as Fire Captain of a Truck Company have been asked a technical question by a new Firefighter and you **DO NOT KNOW** the answer. What action would you take?

POSSIBLE RESPONSE:

In this situation I would inform the Firefighter that I do not know the answer, but that I will get the information and relay it to him.

QUESTION:

You, as Fire Captain, are confronted with **ANIMOSITY** by a particular crew member, what action would you take?

POSSIBLE RESPONSE:

In this situation I would say nothing and try to ascertain the underlying motive.

QUESTION:

Suppose you as Fire Captain had two members of your crew that on several occasions got into **ARGUMENTS** with each other fiercely on the slightest instigation. What action would you take?

POSSIBLE RESPONSE:

In this situation as the supervisor it would be my responsibility to step in and stop these arguments.

QUESTION:

As Fire Captain, how would you handle the situation where one of your Firefighters is invariably screaming and shouting?

POSSIBLE RESPONSE:

In a situation of this nature I would attempt to talk to the Firefighter so as to calm him down.

QUESTION:

Assume as Fire Captain you have an exceptionally productive Firefighter that is constantly **GRIPING** about the work you assign him, how would you handle this situation?

POSSIBLE RESPONSE:

In this situation I would set up a confidential meeting with the Firefighter in order to try and effect a change in attitude on his part.

QUESTION:

As Fire Captain with a new Firefighter in your crew, when is the best time to **CORRECT AN ERROR** made by him concerning station duties?

POSSIBLE RESPONSE:

In this situation the best time to correct the Firefighter is **IMMEDIATELY**.

QUESTION:

As Fire Captain, how would you handle a Firefighter that you have "called on the carpet" for a **POOR PERFORMANCE** of his duties, and the Firefighter tells you that his recent behavior is the result of serious family problems?

POSSIBLE RESPONSE:

In this situation I would suggest to the Firefighter of several social agencies which may be able to help him.

QUESTION:

How would you handle a situation in which a fireman is consistently **MISUSING** equipment?

POSSIBLE RESPONSE:

If I were the **FIRE CAPTAIN** on a piece of apparatus that a fireman was consistently **MISUSING** equipment, I would:

1. Confront the fireman, and advise him as to the proper use of the piece of equipment.
2. If the fireman continues to **MISUSE** the equipment, I would inform him that this misuse of equipment will reflect on his job evaluation and that this information will be passed up through the proper channels.

ADD OR DELETE WHATEVER YOU FEEL IS APPROPRIATE

QUESTION:

You as Fire Captain find that one of your Firefighters work is constantly **BELOW** a reasonable standard. What would be your first course of action?

POSSIBLE RESPONSE:

In this situation it would be best to determine and correct the cause of the sub-standard work performance.

QUESTION:

As Fire Captain one of your Firefighters objects to your **EVALUATION** of his work, how would you handle this situation?

POSSIBLE RESPONSE:

In this situation I would examine the particulars with the Firefighter so that I may show him where improvement may be possible.

QUESTION:

Assume that you as the Fire Captain receive a suggestion from one of your Firefighters, for improving the method of work in the fire station, you **DO NOT** believe that the suggestion is a good one. How would you handle this situation?

POSSIBLE RESPONSE:

In this situation I would discuss with the Firefighter as to why I think his suggestion will not work and at the same time commend him for his job interest.

QUESTION:

What is your opinion of a Fire Captain that tries to delegate **AUTHORITY**?

POSSIBLE RESPONSE:

A Fire Captain cannot delegate **AUTHORITY**.

QUESTION:

As Fire Captain How would you handle a situation where you gave detailed instructions to one of your Firefighters and later you discover that the Firefighter **DID NOT** start the assignment because he did not completely comprehend your instructions?

POSSIBLE RESPONSE:

In this situation I would explain the assignment again, and if feasible illustrate how the assignment is to be accomplished.

QUESTION:

As Fire Captain you give a Firefighter an assignment that he **FAILS** to carry out stating that he misunderstood the orders. How would you handle this situation?

POSSIBLE RESPONSE

In this situation I would take the Firefighters word for it and see if it happens again.

QUESTION:

As Fire Captain you find that one of your Firefighters is **NOT PERFORMING** his duties the way that you requested him to do it, what would be your first response to the Firefighter?

POSSIBLE RESPONSE:

In this situation, the first thing that I would tell the Firefighter is why I want the job performed in this particular manner.

QUESTION:

What would be your method for delegating **UNDESIRABLE** or tiring duties?

POSSIBLE RESPONSE:

I would delegate them using a rotation system.

QUESTION:

What is your opinion of a Fire Captain that delegates most of his **SMALL** task and details?

POSSIBLE RESPONSE:

I think that a good supervising Fire Captain is one that will delegate small task and details.

QUESTION:

As Fire Captain you have a crew member take inventory of some maintenance supplies. The crew member gives you a figure that you feel is **TOO HIGH**. What would be your course of action?

POSSIBLE RESPONSE:

In this situation I would ask the crew member to take inventory over again and explain to him why.

QUESTION:

As Fire Captain of an engine company you get a Firefighter that has **PREVIOUSLY** been assigned to a salvage company. What is the first thing that you should do?

POSSIBLE RESPONSE:

In this situation I would find out how much the Firefighter knows about engine company operations.

QUESTION:

As Fire Captain you find out that your crew members are **NOT** following your instructions, what would be your course of action?

POSSIBLE RESPONSE:

In this situation I would not take any action until I have determined the reason as to why the instructions are not being followed.

QUESTION:

As Fire Captain, what would be your response to a Firefighter that has **SUCCESSFULLY** completed a very difficult task, why?

POSSIBLE RESPONSE:

In this situation I would respond with approval to the Firefighter, because this would stimulate him to strive for increased accomplishment from appreciation of completed assignments.

QUESTION:

If you felt that your supervisor is issuing the **WRONG ORDERS**, how would you confront him?

POSSIBLE RESPONSE:

If I felt that my supervisor was issuing the **WRONG ORDER**, I would:

1. If at a non-emergency, I would discuss the orders with the him.
2. If at an emergency, and the orders would not endanger lives or property, I would follow the orders and discuss the orders with him at a later time.
3. If at an emergency, and the orders would endanger lives or property, I would inform him as to why I thought his orders were inappropriate. I would not endanger lives or property in this situation.

ADD OR DELETE WHATEVER YOU FEEL IS APPROPRIATE

QUESTION:

As Fire Captain, what is the best way to get immediate, **PRODUCTIVE** effort from your Firefighter at a fire?

POSSIBLE RESPONSE:

In this situation I would command the Firefighters in a ethical and dignified manner, without the use of offensive language.

QUESTION:

While explaining a new procedure to your crew, a crew member ask a question about the procedure that you **CANNOT** answer. What action would you take as a Fire Captain?

POSSIBLE RESPONSE:

In this situation I would admit that I don't know the answer and commit to getting the information.

QUESTION:

Assume that you are a Fire Captain conducting a classroom drill and you want your crew to **RETAIN** a particularly essential idea. How would you help your crew remember?

POSSIBLE RESPONSE:

In this situation I would reiterate the information several times during the course of the drill.

QUESTION:

As Fire Captain, what are some of the steps that you would take in breaking down a complicated job into **SMALLER TASK** for your crew?

POSSIBLE RESPONSE:

In this situation I would:

1. Inform each crew member of the importance of his task for the total success of the job.
2. Institute a specific line of work flow and responsibility.
3. Make available written directions for performing each task

QUESTION:

Can you tell us the principle elements of **LEARNING**?

POSSIBLE RESPONSE:

The principle elements of **LEARNING** are: appeal to two or more of senses, repetition, and association.

QUESTION:

What is the primary purpose of **DISCIPLINE**?

POSSIBLE RESPONSE:

The primary purpose of **DISCIPLINE** in the Fire Service is to train the individual.

QUESTION:

Assume that you are a newly promoted Fire Captain and you discover that a subordinate with more time on the Fire Department than you **RESENTS** you. What would be your method of winning respect and cooperation of this individual

POSSIBLE RESPONSE:

In this situation I would make an effort to consult with this individual on details that he has had a number of years of experience with.

QUESTION:

As Fire Captain, how would you **RESOLVE** an argument or confrontation between two of your Firefighters?

POSSIBLE RESPONSE:

In a situation of this type I would bring both Firefighters in for a consultation at the same time and make a resolution in front of both of them.

QUESTION:

As Fire Captain you are giving orders to your crew and one of your Firefighters makes an **UNDESIRABLE REMARK**. This is not the first time that this particular Firefighter has done this. What would be your course of action?

POSSIBLE RESPONSE:

In this situation I would tell the Firefighter that I want to see him as soon as the crew is dismissed.

QUESTION:

What is the most frequent cause for employee **GRIEVANCES**?

POSSIBLE RESPONSE:

The most frequent reason for employee **GRIEVANCES** is wages/salaries.

QUESTION:

As Fire Lieutenant one your Firefighters complains to you that many of the men **HATE** their Drill Instructor. How would you handle this situation?

POSSIBLE RESPONSE:

In this situation I would discourage the Firefighter from spreading "gossip"

QUESTION:

As Fire Captain you are faced with settling a **CRITICISM** from a citizen concerning one of your crew. What would be your first action?

POSSIBLE RESPONSE:

In this situation the first thing that I would do is listen to both the citizen and the crew member in order to try and get to the truth.

QUESTION:

Assume that as Fire Captain you are submitted a **PERSONAL COMPLAINT** by one of your Firefighters. How would you go about achieving an competent and enduring solution?

POSSIBLE RESPONSE:

In this situation I would research the complaint completely and try to discover the underlying cause.

QUESTION:

What is the title usually associated with the person that assist/drives a Chief?

POSSIBLE RESPONSE:

This person is usually called the **CHIEFS AIDE**.

QUESTION:

Assume that you are the acting Battalion Chief and one of your Fire Captains goes directly to the Fire Chief on a matter which he **SHOULD** have consulted with a Battalion Chief first. What action would you take?

POSSIBLE RESPONSE:

In most situations of this nature I would wait until the matter is brought officially to my attention.

QUESTION:

Assume that you are the Battalion Chief and one of your subordinate officers brings you a problem in which he has **NO** solution or recommendations, how would you handle this matter?

POSSIBLE RESPONSE:

In this situation I would suggest to the man that he review the problem and that he should then recommend possible actions for solving the problem.

QUESTION:

Assume that you are a Battalion Chief and one of your subordinate officers that has a lot of support from the men, **DISAGREES** with you on a specific matter concerning Fire Administration. How would you handle this situation and why?

POSSIBLE RESPONSE:

In this situation I would try to win over to my way of thinking so as to use his leadership to an advantage.

QUESTION:

Assume that you are the Battalion Chief of a division that has four new Firefighters that have been assigned to fireboats for six months and never been assigned in the field. What do you feel would be the **BEST ASSIGNMENT** for these Firefighter so that they may perform productively, why?

POSSIBLE RESPONSE:

In this situation I would assign each of the Firefighters to diverse assignments in the field so as to expand their experience.

QUESTION:

Assume that you have been a Fire Lieutenant with a company, but acting as Fire Captain. A new Fire Captain is promoted and assigned to your company. You feel that you are far more **COMPETENT** than the new officer is. What would be the most desirable action for you to take?

POSSIBLE RESPONSE:

In this situation it would be best for me to point out, in the planning stages where I disagree, but carry out the new Fire Captain's decisions once they are made.

QUESTION:

As Fire Captain, how would you deal with a situation where a Firefighter requested a grievance that you know will **NOT** be allowed?

POSSIBLE RESPONSE:

In this situation I would explain to the Firefighter as to the reason that his grievance request will not be allowed.

QUESTION:

As a Fire Captain, how can you best **IMPROVE** the effect when it is necessary to criticize a Firefighter's work?

POSSIBLE RESPONSE:

In this situation, I would mention to the Fire- fighter his good points as well as his weak points.

QUESTION:

As Fire Captain, what would be your first step if you have a probationary Firefighter that consistently performs his assigned task **BELOW** a reasonable standard?

POSSIBLE RESPONSE:

In this situation I would make a decision as to whether or not the reason for his sub-par performance can be corrected.

QUESTION:

As Fire Captain you have a Firefighter that has repeatedly **IGNORED** safety standards. You have been unable to change his behavior. You are irritated by his indifference to the situation. How would you handle this situation if you were concerned about losing your composure?

POSSIBLE RESPONSE:

In this situation it would be best for me to inform the Firefighter that I am annoyed with him and be more firm with him.

QUESTION:

As Fire Captain, what course of action would you take if a Police Officer places one of your Firefighters under **ARREST**?

POSSIBLE RESPONSE:

In this situation, I would suspend the Firefighter and inform the Battalion Chief of the situation.

QUESTION:

How should rules of conduct be given to **PERSONNEL**?

POSSIBLE RESPONSE:

They should be communicated/written clearly and complete to individuals.

QUESTION:

In your opinion, what is the principal responsibility of a Fire Captain in regards to **DISCIPLINE** of crew members?

POSSIBLE RESPONSE:

In these type of situations it is a primary obligation of the Fire captain to discover all of the details concerning each particular matter.

QUESTION:

Assume that you are an acting Battalion Chief observing a drill. You notice that a Firefighter is **NOT** taking an active interest in the various procedures of the drill. What action will you take?

POSSIBLE RESPONSE:

In this situation I would advise the company Fire Captain that he determine the cause for the Firefighters attitude.

QUESTION:

As Fire Captain, you are asked by your Battalion Chief to rate the "**ESPRIT DE CORPS**" of your Firefighters. How would you set up a standard for measuring this level?

POSSIBLE RESPONSE:

In this situation the best standard that I could use would be the consistent aptitude of the crew to produce results for various task for a certain period of time.

QUESTION:

Why is it a good idea for a Fire Captain to **DELEGATE** assignments?

POSSIBLE RESPONSE:

Because he usually is responsible for more functions than he can accomplish himself.

QUESTION:

As Fire Captain you are assigned a job by your Battalion Chief and then you are called away for another task. You leave one of your Firefighters in charge of completing the first job. When you return the Battalion Chief is there and the job is not complete. The Battalion Chief wants an explanation! What will be your **EXCUSE**?

POSSIBLE RESPONSE:

In this situation I would except the responsibility.

RESPONDING

QUESTION:

What would you do if you were responding on a run and your engine **OIL** pressure dropped to zero, no pressure?

POSSIBLE RESPONSE:

If my engine **OIL** pressure drooped to zero while responding or at any time, I would stop as soon as it is safe to do so.

QUESTION:

You are told to respond as a member of a task force to a major brush fire about 100 miles away. You are ordered to respond in an apparatus that you think is obviously **UNSAFE** and unable to make the trip. How would you handle this situation?

POSSIBLE RESPONSE:

I would inform my supervisor as to why I feel that the apparatus is **UNSAFE** and suggest what I thought could be done to make the apparatus safe and able to make the trip, if possible. Under no circumstances would I endanger lives or property in this situation. **ADD WHAT YOU FEEL IS APPROPRIATE FOR YOUR DEPARTMENT**

QUESTION:

What if you were involved in an **ACCIDENT** while driving a fire department vehicle, what is you departments policy and procedure?

POSSIBLE RESPONSE:

If I were involved in an **ACCIDENT** with a fire department vehicle:

1. Stop as soon as safe.(if code 3 notify dispatch to respond another unit to response).
2. Inform my immediate supervisor and the police.
3. Fill out proper paper work
4. Add your departments policy and procedure.

QUESTION:

While you are maneuvering your apparatus at an emergency scene, you accidentally hit a parked car. **NO ONE SAW YOU** hit the car! what will you do?

POSSIBLE RESPONSE:

If I were involved in **ANY** accident with the fire apparatus, I would:

1. Inform my immediate supervisor.
2 Notify the police.
3 Fill out the appropriate paper work.
4 Add your departments policy and procedures.

QUESTION:

You are riding down the street and notice that a gasoline tanker truck is **LEAKING** quite a bit of gasoline, what would you do?

POSSIBLE RESPONSE:

I would try to do the following:

1. Stop the tanker.
2. Notify dispatch to respond the appropriate fire and police units.
3. Evacuate the area.
4. Dike the leaking gasoline.

ADD OR DELETE WHAT YOU FEEL IS APPROPRIATE

QUESTION:

As **FIRE CAPTAIN** of an engine company which is in service on the air while returning from a previous assignment, you hear a call for a second alarm on the radio which you believe will include your company, what action will you take?

POSSIBLE RESPONSE:

I would continue returning toward the station unless my company is dispatched to the incident.

QUESTION:

Assume that you are the **FIRE CAPTAIN** of a truck company and you are ordered, by the **CHIEF** of the department, to respond with your company to correct a hazardous condition at once. What would be your first step in this situation?

POSSIBLE RESPONSE:

Acknowledge the order, and notify dispatch that my particular company is responding.

PERSONAL

QUESTION:

What would you do on your **FIRST SHIFT** assigned as **FIRE CAPTAIN** ?

POSSIBLE RESPONSE:

On my **FIRST SHIFT** as **FIRE CAPTAIN** I would fill the position the same as I would during any other shift as **FIRE CAPTAIN**. That is to fill the position according to the positions responsibilities and requirements. BE PREPARED TO TELL THE ORAL BOARD THE REQUIREMENTS AND RESPONSIBILITIES FOR THE POSITION OF FIRE CAPTAIN!

QUESTION:

If you were going to order a **NEW** pumper or ladder truck, what would you add or delete to the new apparatus compared to the apparatus that you now have in service?

POSSIBLE RESPONSE:

You might consider **NFPA** standards and safety features. You will have to make a list that is appropriate for **YOUR** particular department.

QUESTION:

While you are involved in an assignment at a fire, the **FIRE CHIEF** approaches you with a staggering gate, slurred speech, and watering eyes. He begins to give you obviously inappropriate orders, what would you do?

POSSIBLE RESPONSE:

I would inform the **FIRE CHIEF** that I am already responsible for an assignment. I would then inform the **FIRE CHIEF** that he was looking ill and recommend that he be checked out by the paramedics.

ADD OR DELETE WHATEVER YOU FEEL IS APPROPRIATE

QUESTION:

After a fire, you are standing next to your apparatus while you are still on the scene. A thankful property owner **OFFERS** you a beautiful antique model fire engine, worth several hundred dollars.You at first refuse but the fellow insists, saying that he was just going to dispose of it anyway. He then places it in the captains seat. What would you do?

POSSIBLE RESPONSE:

I would inform my supervisor of the situation and let him handle the situation through the **PROPER CHANNELS**
ADD/DELETE WHAT IS APPROPRIATE FOR YOUR DEPARTMENT

QUESTION:

You are eating dinner at a restaurant and notice that the owner has **LOCKED** the required exits. What would you do?

POSSIBLE RESPONSE:

I would inform the restaurant manager or owner about the ordinance and of the dangers. I would have him **UNLOCK** the exit door and inform the proper authorities.

QUESTION:

You are performing overhaul and **OBSERVE** another firefighter putting something into his turnout pocket, what would you do?

POSSIBLE RESPONSE:

I would confront the firefighters and **GET THE FACTS** as to what and why.

QUESTION:

As Fire Captain you find that you have reprimanded, unjustly, a Firefighter for an incident that took place on a previous response. How would you handle this situation?

POSSIBLE RESPONSE:

In this situation I would immediately admit my error to the Firefighter.

QUESTION:

Give some examples of **NEGATIVE DISCIPLINE**.

POSSIBLE RESPONSE:

Some examples of **NEGATIVE DISCIPLINE** are:

1. Suspension.
2. Discharge.
3. A "chewing out".

FIREFIGHTING

QUESTION:

As a **FIRE CAPTAIN**, what are you thinking about while responding to a fire?

POSSIBLE RESPONSE:

As a **FIRE CAPTAIN**, some of the things that I would be thinking about are:

1. The response location.
2. The response route.
3. Topography of area.
4. Access and alternate routes.
5. The response routes of other responding units.
6. Hydrant locations.
7. Fire flow.
8. The quality of the water supply in the area that I am responding to.
9. If the fire building is sprinklered.
10. If the fire building has standpipes.
11. If the fire building has any special fire fighting devices.
12. If the business in the fire building has any private fire fighting facilities or firefighters.
13. The time of day:
    - Visibility
    - Traffic conditions
14. Weather conditions:
    - Visibility
    - Road conditions
15. Operation of apparatus.
16. Fire load of the fire building.
17. Exposures in the area of the fire building.
18. Physical location of fire building.
19 Occupancy classification.
20 Occupant load.
21. Structural condition.(internal and external)
22. Anticipated fire behavior from pre-plan attack.
23. Public attitude in particular area.
24. Predetermined objectives which can be adjusted.

QUESTION:

As a **FIRE CAPTAIN**, what would you be thinking about at a fire scene?

POSSIBLE RESPONSE:

As a **FIRE CAPTAIN**, some of the things that I would be thinking about, at a fire scene, are:

1. The water supply.
2. Amount of hose in use.
3. Apparatus location.
4. Where responding units are coming from.
5. Radio communications.
6. Safety of personnel.
7. Proper use of apparatus and equipment.
8. Command the incident until relieved by a superior officer.
9. Merging of assignments with other companies so as to minimize efforts towards the same goals.
10. Be aware of:
    - Amount of water being used and available.
    - Other pumpers and apparatus on scene.
    - Other pumpers and apparatus responding.
    - Equipment available.
    - Radio communications.
    - Location of hose lines:
        - Inside and outside.
        - Ground floors and basements.
        - Upper elevations.

ADD ANY OF YOUR OWN APPROPRIATE IDEAS

QUESTION:

In **RELAY** operations, where should the largest pumper be in relation to the water source?

POSSIBLE RESPONSE:

In **RELAY** operations, the pumper with the largest capacity should be the pumper nearest the water source.

QUESTION:

What would you do if the standpipe connection on a standpipe system was **DAMAGED**, so as not to allow you to connect supply lines to them

POSSIBLE RESPONSE:

I would connect the supply line to the first floor that has **DISCHARGE GATES** (second floor), with appropriate amount of hose with a double female fitting.

QUESTION:

In laying **DUAL LINES** to a sprinkler system you end up 100 feet short, how would you handle this situation?

POSSIBLE RESPONSE:

I would disconnect 100 feet from one supply line and attach it on to the other supply line, then charge the sprinkler system with one line until the additional amount of hose (200 feet) could be put into service.

QUESTION:

If you are in a burning building and have advanced as far as the heat and smoke permit without actually seeing flames. What is the best action to take?

POSSIBLE RESPONSE:

In this situation the best action to take would be to **VENTILATE** that section of the building if possible, and then advance further.

QUESTION:

As **FIRST IN OFFICER** you are aware that the initial **POLICE** response is inadequate to control the situation. What would be your course of action?

POSSIBLE RESPONSE:

I would ask for additional **POLICE** support.

QUESTION:

Assume that you are the **FIRE OFFICER** in charge of a fire in a large building that has smoke coming out of the eaves, cracks around doors and windows, and other openings in the building. This situation should signal to you the possibility of **BACKDRAFT**. What action could you take under these conditions to reduce the possibility of **BACKDRAFT**?

POSSIBLE RESPONSE:

Under these circumstances I would get my lines in position and **VENT** the gases from the roof.

QUESTION:

Assume that you are the first in **FIRE ENGINE COMPANY** at a fire in a commercial building. Give us an example of when you would not advance the first hose line into the building.

POSSIBLE RESPONSE:

It would not be advisable to advance the first in hose line into the building when the fire is beginning to extend to **EXPOSURES**.

QUESTION:

As **FIRST IN OFFICER** to a fire within a row of one-story stores, with smoke making it difficult to locate the exact fire location. What would be a good method of locating the fire origin immediately upon your arrival?

POSSIBLE RESPONSE:

I could look to see if one of the store front windows has been discolored by the heat.

QUESTION:

Assume that you are the **FIRST IN OFFICER** at the scene of a fire, unless the situation obviously calls for other action, what is generally conceded that your first action should be directed towards?

POSSIBLE RESPONSE:

In this situation, my first action should be directed toward determining whether or not lives are endangered.

QUESTION:

You are the **OFFICER IN CHARGE** of an engine company dispatched to check an odor of smoke in a three level, 24 unit, garden apartment complex. It is a wood frame brick veneer structure and has no sprinkler system. The complex does have an operational fire alarm system, master connected to the Fire Department with smoke detectors in the hallways, heat detectors in all apartments and pull stations on every level and at all exits. The alarm system however has not been activated.

When you arrive you detect no visible smoke outside the building, however, when you enter the building you are met by a woman who claims she is the one who called the Fire Department. On your way to check out her apartment you find a very light smoke condition in the second floor hallway, when you reach the open door of her apartment you find the smoke condition worse than the hallway but it is still light smoke. It is 1100 hours on a Friday.

As you might expect conditions inside the structure worsen. One of your men inform you that while investigating the smoke odor on the first floor he found a plumber smoldering some pipes in one of the apartments below and across the hall from the callers apartment. At about the same time you find the source of the smoke you are looking for. The source is an electrical outlet in the bathroom of the callers second floor apartment. To make matters worse the building's fire alarm system begins to sound.

With these conditions in mind tell us why you think this situation has evolved.

POSSIBLE RESPONSE:

A good knowledge of building construction and fire behavior are essential tools for the Incident Commander (I/C) as well as

Company Officers. A proper size-up is almost impossible without some understanding of the specific qualities of the structure you are dealing with. Many of the decisions a fire officer will make are based upon a combination of the present conditions of an incident and what he anticipates will be happening in three or five or even ten minutes from the present. Fire load, fire extension, building materials, occupancy, point of origin all have an enormous impact on what the fire officers next move will be. Will he call for more manpower, another ladder company, another engine or maybe conditions are not as bad as they look.

Fires in garden apartment complex's can be difficult at best because of some of their special construction characteristics. Many of these buildings were built in the middle sixties's to the early seventies. They were being constructed faster than appropriate fire codes could be adopted. The result was that many of these structures were lost before firefighters were able to establish procedures to combat these fires and before appropriate fire codes could be established and enforced.

One of the more common fire problems with this type of construction is a plumber smoldering pipes in a bathroom or kitchen of an apartment and starting a fire in the wall. In most cases the water and drain pipes for the bathrooms and kitchens were contained in common channels or pipe cases. These pipe cases allow for horizontal as well as vertical spread of fire and heat and can remain in the walls for sometime without being discovered, sometimes jumping over hallways, floors and apartments before anyone detects a problem.

QUESTION:

Assume that you are the **FIRST IN OFFICER** at a fire where persons require **RESCUING**, what course of action will you consider taking first?

POSSIBLE RESPONSE:

In this situation, it is best to consider the possibility of delaying the **RESCUES** until other factors are studied.

QUESTION:

Assume that you are the **FIRST IN OFFICER** at an apartment fire and you observe flames at a front window of the second story. It is apparent that the fire originated close to this window and that a small portion of the floor is involved. What action will you take? Why?

POSSIBLE RESPONSE:

In this situation, I would attack the fire from the inside drawing the hose-line up the interior stairs. In such instances the more efficient attack is from the inside of the structure primarily because fire streams can be pointed directly at the fire.

QUESTION:

What is the **MOST EFFECTIVE** means of protecting adjacent structures from an exposure fire across a small open space such as an alley?

POSSIBLE RESPONSE:

In this situation the most effective means would be by the use of heavy fire streams.

QUESTION:

Assume that you are the **FIRST IN OFFICER** at a loft fire with fire lapping out of the upper floor windows. Your initial attack was with a master fire stream to prevent the spread of fire. What would be the main reason for shutting down the master stream as soon as hand-lines are in place on the stairway?

POSSIBLE RESPONSE:

The main reason for shutting down the master stream in this situation would be to avoid driving smoke down into the stairway.

QUESTION:

As **FIRST IN OFFICER** to a fire that has spread to the stairs, elevators and corridors of an old non-fireproof hotel, as far as life hazard, where would your greatest concern be directed to?

POSSIBLE RESPONSE:

In this situation my greatest concern as far as life hazards are concerned would be directed to the top floor.

QUESTION:

Assume that your company has been fighting a fire, together with several other companies, for a long period of time. As **FIRE CAPTAIN** what would be your course of action if your fuel supply on your pumper is running low?

POSSIBLE RESPONSE:

In this situation, I would notify the officer in charge that my fuel supply was running low.

QUESTION:

What **RESULT** would you expect from placing a charged fog nozzle into a closed room that is on fire?

POSSIBLE RESPONSE

I would expect that the water would turn to steam and smother the fire.

QUESTION:

Upon arriving at a fire you find that a single family **DWELLING** is involved. Dense smoke is pouring out of the open windows at the front of the house and is being forced out around the front door. Before ordering the opening of the front door, what would be your course of action?

POSSIBLE RESPONSE:

Before ordering the opening of the front door, I would get the hose lines into position at the front door.

QUESTION:

Assume that you are the **FIRST IN OFFICER** to a fire involving a vacant building where conditions exist which permit the implementation of an exterior attack. What would be your initial tactic?

POSSIBLE RESPONSE:

In this situation, my initial tactic would be the use of heavy fire streams.

QUESTION:

As a **FIRE OFFICER** confronted with the problem of major exposure hazards from an active fire, how can you determine if an exposed wall contains excessive heat?

POSSIBLE RESPONSE:

In this situation you can tell if an exterior wall contains excessive heat by sweeping the wall with a fog spray, if the wall has excessive heat it will be indicated by a cloud of condensed steam.

QUESTION:

Assume that at a fire scene you have been directed to **CUT OFF** the gas to the fire valve and you do not know the location of the valve. To what area would you go to first?

POSSIBLE RESPONSE:

In this situation, I would go just outside the building wall.

QUESTION:

What would be the safest and most effective procedure for **SHUTTING DOWN** the low pressure gas supply to an industrial complex?

POSSIBLE RESPONSE:

In this situation the method to use is first shut off the gas meter valve, then shut off the main service valve. You should also check for the presence of a curb valve and, if there is one, shut it off also.

QUESTION:

Assume you are the **FIRE OFFICER** in charge of a fire in a commercial building that has a reservoir containing 10% **ACETIC ACID**. How would this condition effect your course of action?

POSSIBLE RESPONSE:

In this situation, I would operate as though the condition did not exist.

QUESTION:

As a **FIRE CAPTAIN**, what would be your greatest concern in a fire involving bleaching powder?

POSSIBLE RESPONSE:

In this situation my greatest concern would be a toxic and irritant gas.

QUESTION:

Assume that you as the **FIRE CAPTAIN** working with your crew in the basement of a **DRUG STORE** fire and a large stock of chemicals are stored on open shelves in wooden boxes. What precautions should you and your crew take if you can see flames, through thick smoke, behind a row of boxes?

POSSIBLE RESPONSE:

In this situation, we should avoid upsetting and mixing the chemicals, which could possibly cause an explosion.

QUESTION:

Assume that you are the **FIRE OFFICER** in charge of a fire in a warehouse that contains several cans of calcium carbide for shipment to the Nova. The fire has reached the cans, but it is not known if any of the cans have ruptured. What method would you use to control this fire?

POSSIBLE RESPONSE:

In this situation, I would ventilate and fight the fire with water streams as for most any other fire.

QUESTION:

What would be your course of action at a large tank fire leaking **HYDROCHLORIC ACID** on a metal surface? Why?

POSSIBLE RESPONSE:

In this situation, I would want to neutralize the acid with a material such as soda ash to prevent the possibility of an **EXPLOSION** hazard by the action of the acid on metal materials.

QUESTION:

As the **FIRST IN FIRE OFFICER** at a heavily involved supermarket fire, during heavy shopping hours, where would you order your first in hose lines?Why?

POSSIBLE RESPONSE:

In this situation, it would be appropriate order the hose lines through the rear of the structure, because it would create less interference with the escape routes of the occupants.

QUESTION:

Considering hazards involved and the effectiveness, what is the most practical method of the use of an **INDIRECT FIRE ATTACK**?

POSSIBLE RESPONSE:

Considering the hazards involved and the effectiveness of an **INDIRECT FIRE ATTACK**, the most practical method is to make the **INDIRECT FIRE ATTACK** from outside the structure.

QUESTION:

As **FIRST IN OFFICER** to an incident involving a **CHLORINE** leak, when persons must be evacuated, what type of an area would you move these people to?

POSSIBLE RESPONSE:

In this situation, it would be best to guide these people to an area above the level of the leak and to the windward side of the leak.

QUESTION:

Assume that you are the **FIRE CAPTAIN** of the first in engine company at the scene of a fire. After you order the appropriate hose stretch, how will you determine the most practical point where surplus hose should be located?

POSSIBLE RESPONSE:

In this situation, the surplus hose should be located in an area that will allow smooth advancement of the hose-line while attacking the fire.

QUESTION:

What are the two objectives that **FIRE EXTINGUISHMENT** depends upon in fighting unconfined building fires?

POSSIBLE RESPONSE:

The two objectives are:

1. The rate at which heat is transmitted to the water being utilized must surpass the rate of creation minus the natural expenditure of heat by radiation and convection to the exterior.
2. Residual heat must be decreased to a point which will allow firefighting personnel to enter and extinguish the remaining fire.

QUESTION:

Assume that you are the **OFFICER IN CHARGE** of an incident involving the rupture in the process piping of an **LNG** storage installation resulting in a gas or liquid leak that does not become involved in fire. What would be your course of action and why?

POSSIBLE RESPONSE:

In this situation it is recommended to use water spray on the fuel, because the presence of the water spray will dilute the oxygen content in the vapor-air mixture, thus reducing the possibility of an explosion.

QUESTION:

As **FIRST IN OFFICER** at a building where **DUST EXPLOSION** is a hazard, you should first order?

POSSIBLE RESPONSE:

In this situation, a low velocity fog stream should be used first.

QUESTION:

Assume that you are the **FIRST IN OFFICER** to the scene of an **AMMONIA** leak, in a large industrial complex that has a diffuser in the system, what would be your course of action?

POSSIBLE RESPONSE:

In this situation it would be best to get water flowing in the diffuser before the **AMMONIA** is allowed to enter.

QUESTION:

Assume that it is necessary to enter an un-ignited **LPG** cloud in order to effect a shut-off, what is the best course of action?

POSSIBLE RESPONSE:

In this situation, it is best to don breathing apparatus and move towards the shut-off area with the use of fog streams to dissipate the vapor.

QUESTION:

As **FIRE OFFICER** in charge of an **OIL FIRE** of a large proportion, how would you recommend extinguishment of this fire?

POSSIBLE RESPONSE:

In this situation, it is recommended to direct a stream of foam lightly on the surface of the oil.

QUESTION:

Assume that you are the **FIRE OFFICER** in charge of an incident involving a tank containing **BUTANE**, with a broken fitting that is allowing the liquid to escape and feed the fire, what would be your course of action in combatting this fire?

POSSIBLE RESPONSE:

In this situation, it is advisable to keep the tank cooled by water streams, cover exposures, and permit the fire to burn itself out.

QUESTION:

Assume that you are the **FIRST IN OFFICER** at an incident involving a shipment belonging to the **ATOMIC ENERGY COMMISSION**, in relation to fire streams, what would your orders be?

POSSIBLE RESPONSE:

In this situation, it would be best to attack with solid streams at the maximum range.

QUESTION:

As the **FIRE CAPTAIN** first in at an airport, you respond to an incident involving overheated brakes on a large jet, a fire has started on one of the wheels, how would you handle this incident?

POSSIBLE RESPONSE:

In this situation, I would use a dry chemical fire extinguisher and attack the fire from a front and back position rather than from the side of the wheel.

QUESTION:

Assume, during a mild windless day, you are the **FIRE OFFICER** in charge of an incident involving the collision of a **GASOLINE** tank truck, with considerable **GASOLINE** spilled on the street, after the fire is extinguished what should be done to the **GASOLINE**?

POSSIBLE RESPONSE:

In this situation, the **GASOLINE** should be covered with foam.

QUESTION:

If a **GASOLINE** tank truck is spilling **GASOLINE**, where is the best place for you to have your Firefighter look first for the emergency valve or device for shutting down the flow of **GASOLINE**?

POSSIBLE RESPONSE:

In this situation, it is best to have the Firefighter look at the end of the truck opposite the regular discharge outlets.

QUESTION:

What is a quick method which will help prevent glass windows from cracking at an incident involving a large ignited **GASOLINE** spill near a building with large plate glass windows?

POSSIBLE RESPONSE:

In this situation, it will help to apply foam directly to the windows.

QUESTION:

As **FIRE CAPTAIN** in charge of a **GASOLINE** fire occurring in a **SPRINKLERED** garage, what would you expect the water from the sprinklers to do?

POSSIBLE RESPONSE:

In this situation, I would expect the water from the sprinklers to cool the surrounding area and limit the spread of the fire.

QUESTION:

Assume that you are the **FIRST IN OFFICER** on a first alarm to a fully involved four story apartment complex at 4:00 AM. The fire is seriously threatening adjoining dwellings. What should your first course of action be?

POSSIBLE RESPONSE:

In this situation, I would order the evacuation of occupants and ask for additional help.

QUESTION:

Assume that you are fighting a fire on the roof of an eight story building and the line **BURST** between the seventh and eight floor. What would you do in this situation after shutting down the water?

POSSIBLE RESPONSE:

In this situation, after shutting down the water it would be best to disconnect the line at the ground, insert a new section of hose, pull the line to the roof, and remove the burst section of hose.

QUESTION:

Assume that you are the **FIRST IN OFFICER** of a first alarm on a high rise incident. You notice smoke pushing out of the exterior skin of the building at about the 16th floor. What should be your first step?

POSSIBLE RESPONSE:

In this situation, I would call for more help.

QUESTION:

Assume that you are **FIRST IN OFFICER** at a fire involving the eighth floor of a loft building equipped with a standpipe. When you attempt to lay a 2 1/2" line to the standpipe, you find that the siamese connection is damaged and cannot be used. hat is the most desirable way to get water to the fire floor?

POSSIBLE RESPONSE:

In this situation, the most desirable way to get water to the fire floor is to stretch a hose line to the first floor and hook into the standpipe connection on that floor.

QUESTION:

Assume that you are the **FIRE CAPTAIN** in the first arriving engine company at the scene of an emergency involving a twelve story apartment building. The elevator, with center-opening doors, has plunged from the twelfth floor to about the eighth floor level. There are four passengers on board, but they are unable to assist rescue efforts. What would be the most advisable step to take?

POSSIBLE RESPONSE:

In this situation, the most advisable step to take would be to order the power shut-off and try to pass a hook to the hoist-way interlock from the floor above.

QUESTION:

Assume that you are the **FIRE OFFICER** in charge of the second in ladder company at a medium fire situation involving an occupied, detached, private dwelling, around 04:00 hours. What would be your main objective?

POSSIBLE RESPONSE:

In this situation, my main objective would be to increase search and rescue, then assist as needed.

QUESTION:

Assume you are the **FIRE CAPTAIN** fighting a fire on the roof of a building via the outside of the structure. How much hose would you want on the roof?

POSSIBLE RESPONSE:

In this situation, I would want at least two lengths of hose available on the roof.

QUESTION:

Can you explain what the problem is at a **WELL ADVANCED FIRE** in a sprinklered structure, where the sprinklers are in operation with an engine connected to the siamese pumping at a pressure of 100 PSI. Your Engineer advises you that his engine pressure has suddenly fallen drastically, and that he cannot maintain the 100 PSI?

POSSIBLE RESPONSE:

The probable cause for the loss of pressure, in this situation, would be that additional sprinkler heads have opened from the effect of the spread of fire.

QUESTION:

What would be the **MOST LIKELY** reason for the Chief in charge of a building fire to order you as the Captain of a pumper company, to hook a line immediately to the sprinkler siamese of the building adjacent to the fire building?

POSSIBLE RESPONSE:

In this situation the Chief officer is most likely trying to insure water at sufficient pressure to the sprinklers.

QUESTION:

Where should **VENTILATION** be provided in a minor fire with a small amount of flame on the second floor of a eight story chemical warehouse?

POSSIBLE RESPONSE:

In most fires of this type, **VENTILATION** should be provided by opening up at the street level.

QUESTION:

In the following situation you are the Captain of a **LADDER COMPANY**: A pipe supplying **NATURAL GAS** to an eight story apartment building has a large leak in it causing a strong order on the 5th and sixth floors. After you shut off the gas supply, what would be your course of action in regards to **VENTILATION**?

POSSIBLE RESPONSE:

As far as **VENTILATION** in this situation it is best to **CROSS-VENTILATE** all floors 5th and above by natural ventilation and to open the roof.

QUESTION:

What action would you take as the Captain of a **LADDER COMPANY** that, after cutting a hole in the roof of a fire building, you find that only a small draft is taking place?

POSSIBLE RESPONSE:

In this situation, the first action that should take place is to probe the hole with a pike pole or other appropriate tool to check for obstructions.

QUESTION:

In the same situation as the previous question, with the exception that smoke was **ALSO** coming from the window where the victim was located. What would be your first course of action?

POSSIBLE RESPONSE:

In this situation it would be best to order a ladder placed at the window where the victim is located.

QUESTION:

Assume that you are the Captain of an engine company first in on a fire involving a **TRANSFORMER** located between low voltage and a high voltage wires, what would be your course of action?

POSSIBLE RESPONSE:

In this situation it would be best to call the power company, allow the fire to burn it's self out, and stand by until the power company arrives.

QUESTION:

Assume that you are the Captain of the first in engine company to a fire at a two story wood frame dwelling. While sizing-up the situation you observe smoke coming out from many of the upstairs windows. You also observe a **VICTIM** calling for help from another upstairs window, from which no smoke is coming. What would be your first course of action?

POSSIBLE RESPONSE:

In this situation it would be best to send a crew to bring the victim out, using the stairs.

QUESTION:

Assume that you are the Captain in charge of a **LADDER COMPANY** performing **OVERHAULING** operations in a severely damaged structure that has a large water tank on the roof that is half full of water. What would be your first critical step in this situation?

POSSIBLE RESPONSE:

In this situation the first step should be to empty the water tank before initiating the overhaul operations.

QUESTION:

Assume that you are the Captain on a Fireboat that is called upon to assist at a flammable liquid spill (**JP-4**) on the surface of the water. What would be the most appropriate course of action in this situation?

POSSIBLE RESPONSE:

In this situation, if possible, break up the spill with hose streams, or contain the spill with the forward turret.

QUESTION:

Assume that you are the Captain of a pumper company that arrives first at a **VEHICLE** fire which was caused by a short circuit under the hood and has spread to the dash board, the wires are still smoldering, What will be your course of action?

POSSIBLE RESPONSE:

In this situation it would be best to extinguish the fire under the hood and disconnect the battery terminal connections.

QUESTION:

As the Captain of a pumper company, one of your Firefighters reports to you on the fire ground an automobile **DROVE** over your supply line, with the added information that the Firefighter gave the license number and make of the car to the Police Department and they passed the matter up. What course of action would you take?

POSSIBLE RESPONSE:

In this situation it would be best to make a special report to the Chief of the Fire Department.

QUESTION:

Assume that you are the Captain of the first in engine company to a fire in a **HIGH VOLTAGE ELECTRICAL VAULT** with a large amount of smoke and fire coming from the vents. What would be your course of action?

POSSIBLE RESPONSE:

In this situation it would be best to call the utility company, apply dry chemical through the vents, and stand by until the utility company arrives.

QUESTION:

Assume that you are in command at a fire in the hold of a ship carrying a cargo of **AMMONIUM NITRATE**. Your Firefighter informs you that when he directs a solid stream on the piles of the ammonium nitrate, several eruptions with the force of minor explosions take place. In this situation, what would be the best course of action to take?

POSSIBLE RESPONSE:

In this situation, the best course to follow is to increase the volume of water applied to the fire.

SECTION 6

FIRE PREVENTION

QUESTION:

What are the classification of **OCCUPANCIES**:

POSSIBLE RESPONSE:

The classification of **OCCUPANCIES** are:

1. A = Assembly.
2. B = Business.
3. E = Educational.
4. H = Hazardous.
5. I = Institution.
6. M = Carports - fences.
7. R = residences.

QUESTION:

Tell us how your Cities **FIRE PREVENTION** program works.

POSSIBLE RESPONSE:

GIVE THE APPROPRIATE ANSWER FOR YOUR CITY

QUESTION:

Tell us about some of the major **TARGET HAZARDS** within your City.

POSSIBLE RESPONSE:

GIVE EXAMPLES of **TARGET HAZARDS** such as:

1. Buildings with high life hazards.
   A. HOSPITALS.
   B. REST HOMES.
   C. INSTITUTES. etc.
2. Buildings with high potential of property loss.
   A. LUMBER YARDS.
   B. WAREHOUSES.
   C. SHOPPING CENTERS.
   D. MULTIPLE RETAIL OUTLETS.
   E. CHEMICAL HANDLING BUSINESSES. etc.

GIVE SOME PARTICULARS FOR YOUR CITY

QUESTION:

What are some of the **LARGER BUILDINGS** in your City with **SPRINKLER SYSTEM HOOKUPS**, and where are the hookups located?

POSSIBLE RESPONSE:

GIVE EXAMPLES such as:
1. HOSPITALS.
2. WAREHOUSES.
3. INSTITUTIONS. etc.

KNOW THE LOCATION OF SPRINKLER HOOKUPS!

QUESTION:

What is the purpose of a **FIRE WALL**?

POSSIBLE RESPONSE:

The purpose of a **FIRE WALL** is to prevent the spread of fire.

QUESTION:

What is the best way of reducing the fire and life hazards of **PUBLIC** buildings?

POSSIBLE RESPONSE:

The best method for reducing fire and life hazards in public buildings is the use of appropriate fire resistant construction for the occupancy.

QUESTION:

As Fire Captain you discover a building that appears to be about to **COLLAPSE**, what would be your course of action?

POSSIBLE RESPONSE:

In this situation I would notify headquarters and fill out an appropriate official report.

QUESTION:

As Fire Captain of a small City Fire Department what kind of methods could you suggest for **REDUCING** the cost of Fire Protection?

POSSIBLE RESPONSE:

Some of the suggestions that I could give in this situation are:

1. That possibly one Fire Department for a metropolitan area as opposed to several separate Fire Departments in each municipality.
2. Unification of Fire Departments can save large amounts of money, using ISO company and manpower standards.
3. Mutual and automatic aide.

QUESTION:

Although a **CHECK LIST** may be a helpful tool for a Fire Captain to use while conducting Fire Company Fire Prevention inspections, what is the main draw back involved in its use?

POSSIBLE RESPONSE:

The main draw back in the use of a check list is that circumstances that are not covered on the list may be omitted.

QUESTION:

As a Fire Captain conducting a building inspection, how would you determine if some rolls of **FILM** are of the safety type?

POSSIBLE RESPONSE:

In this situation I would take a small piece of the film outside of the building and ignite it.

QUESTION:

As Fire Captain you are making a routine fire inspection of an occupancy and you are approached by a citizen who makes some **UNFAVORABLE** comments about the Fire Chief. What would be your course of action?

POSSIBLE RESPONSE:

In this situation I would try to change the subject as soon as possible.

QUESTION:

A Fire Captain during an investigation for the cause of a fire ask a person **LEADING QUESTION:S**, what is your opinion of this tactic?

POSSIBLE RESPONSE:

This type of questioning is inappropriate primarily because this type of questioning makes the free exchange of information difficult.

QUESTION:

As Fire Captain, you determine that the cause of a spray booth fire was because of **ALTERNATE SPRAYING** of lacquer and enamel, without cleaning the spray booth. Why would this cause the fire?

POSSIBLE RESPONSE:

In this situation the fire would be the result of **SPONTANEOUS** ignition of the residue.

SECTION 7

ELIMINATOR QUESTIONS

QUESTION:

How much **RESERVE** pumping capacity should a pumper have?

POSSIBLE RESPONSE:

Fire pumpers should have a **RESERVE** pumping capacity of **12%** or 1/8th of there rated capacity.

QUESTION:

What are the standards for the **U.L. ACCEPTANCE** test for double jacket fire hose?

POSSIBLE RESPONSE:

U.L. ACCEPTANCE test for double jacket fire hose:

1. Elongate 42 inches or less.
2. No more than 1 3/4 turns to the right. (tightens the couplings)
3. Hydrostatic pressure of 400 PSI.

QUESTION:

What **PSI** is new single jacket hose tested at?

POSSIBLE RESPONSE:

New single jacket hose is tested at **300 PSI**.

QUESTION:

What is the **PRESSURE LOSS** in unlined hose, as compared to lined hose?

POSSIBLE RESPONSE:

The **PRESSURE LOSS** in unlined hose is about two (2) times that of lined hose.

QUESTION:

What happens to friction loss in hose if it is laid out in a **ZIG-ZAG** pattern, compared to if it were in a straight line?

POSSIBLE RESPONSE:

Hose in a **ZIG-ZAG** pattern will increase friction loss by about 5% to 6%.

QUESTION:

What is the **FRICTION LOSS** in two lines of 2 1/2" hose as compared to the friction loss in a single line of 2 1/2" hose?

POSSIBLE RESPONSE:

The **FRICTION LOSS** in two lines of 2 1/2" hose is about **28%** of the **FRICTION LOSS** in a single line of 2 1/2" hose.

QUESTION:

What happens to the **FRICTION LOSS** if a length of hose is doubled?

POSSIBLE RESPONSE:

If the length of hose is doubled the **FRICTION LOSS** is doubled.

QUESTION:

Which **N.F.P.A.** pamphlet is the standard for **FIRE HOSE**?

POSSIBLE RESPONSE:

N.F.P.A. #196 is the standard for **FIRE HOSE**.

QUESTION:

Which **N.F.P.A.** pamphlet is the standard for the **CARE OF FIRE HOSE**?

POSSIBLE RESPONSE:

N.F.P.A. #198 is the standard for the **CARE OF FIRE HOSE.**

QUESTION:

What is the minimum amount of **RESERVE** fire hose that is to be on hand?

POSSIBLE RESPONSE:

The minimum amount of **RESERVE** fire hose on hand is **ONE FULL LOAD.**

QUESTION:

What are the maximum **EFFICIENT** water capacities of different sizes of fire hose?

POSSIBLE RESPONSE:

Maximum **EFFICIENT** water capacities of fire hose are:

1. 1" hose = 30 GPM.
2. 1 1/2" hose = 100 GPM.
3. 2 1/2" hose = 250 GPM.
4. 3" hose = 500 GPM.
5. 3 1/2" hose = 750 GPM.

QUESTION:

As far as **FLOW** is concerned, would 3" hose compare to three 2 1/2" hose lines?

POSSIBLE RESPONSE:

As far as **FLOW** is concerned, two 3" hose lines would equal three 2 1/2" hose lines:
EXAMPLE:
1. 3" line = 375 GPM (**BEST** flowing capacity).
2. 2 1/2" line = 250 GPM (**BEST** flowing capacity).
3. 2 X 375 = 750 GPM.
4. 3 X 250 = 750 GPM.

QUESTION:

How much does a 50 foot section of 2 1/2" cotton rubber lined hose **WEIGH** without water, and not counting the couplings?

POSSIBLE RESPONSE:

50 foot section of single jacket would **WEIGH 35 LBS**
and a 50 foot section of double jacket would **WEIGH 50 LBS**.

QUESTION:

How much does a 50 foot section of 2 1/2" cotton rubber lined hose filled with water **WEIGH**?

POSSIBLE RESPONSE:

A 50 foot section of 2 1/2" cotton rubber lined hose filled with water **weighs about 106 lbs**.(12.75 gallons X 8.35 lbs = 106 lbs).

QUESTION:

What are the **NOMINAL** lengths for fire hose?

POSSIBLE RESPONSE:

The **NOMINAL** lengths for fire hose are:
1. 50 foot length of hose = no less than 48 feet.
2. 75 foot length of hose = no less than 73 feet.
3. 100 foot length of hose = no less than 98 feet.

QUESTION:

As we normally think of it in the fire service, what is **PRESSURE**?

POSSIBLE RESPONSE:

PRESSURE, as we normally think of it in the fire service, is that force delivered by pumpers for supplying water for fire fighting.

PRESSURE may be defined as the measurement of energy contained in water.

QUESTION:

How much does a cubic foot of water **WEIGH**?

POSSIBLE RESPONSE:

A cubic foot of water **WEIGHS** about **62.5 LBS.**

QUESTION:

How many cubic inches does one cubic foot of water **CONTAIN**?

POSSIBLE RESPONSE:

One cubic foot of water **CONTAINS 1728 CUBIC INCHES**.

QUESTION:

How many cubic inches **CONTAINED** in gallon of water?

POSSIBLE RESPONSE:

One gallon of water **CONTAINS 231 CUBIC INCHES**.

QUESTION:

How much does one gallon of water **WEIGH**?

POSSIBLE RESPONSE:

One gallon of water **WEIGHS 8.35 LBS**.

QUESTION:

How many gallons **CONTAINED** in a cubic foot?

POSSIBLE RESPONSE:

There are **7.481 GALLONS CONTAINED** in a cubic foot.

QUESTION:

How much does a column of water 1' high and 1" square **WEIGH**?

POSSIBLE RESPONSE:

A column of water 1' high and 1" square **WEIGHS** 4.34 LBS.
(note .5 PSI for field hydraulics)

QUESTION:

How much PSI will a column of water **2.304 FEET** in height **EXERT** at its base?

POSSIBLE RESPONSE:

A column of water **2.304 FEET** will **EXERT 1 Lb PSI** at its base.

QUESTION:

What does **PSIG** stand for?

POSSIBLE RESPONSE:

PSIG = pounds per square inch gaged in excess of normal atmospheric pressure.

QUESTION:

What does **PSIA** refer to?

POSSIBLE RESPONSE:

PSIA = absolute pressure; pressure above zero.

QUESTION:

What is **ABSOLUTE ZERO PRESSURE**?

POSSIBLE RESPONSE:

ABSOLUTE ZERO PRESSURE = a perfect vacuum; a complete absence of pressure.

QUESTION:

Will a **HARD SUCTION** hose give more supply than a soft suction hose from a hydrant.

POSSIBLE RESPONSE:

HARD SUCTION from a hydrant will give less than 10% more supply than a soft suction.

QUESTION:

At what temperature is water considered to be at **MAXIMUM DENSITY**?

POSSIBLE RESPONSE:

MAXIMUM DENSITY of water is at 39 degrees F.

QUESTION:

What is **ATMOSPHERIC PRESSURE** at sea level, and how does altitude affect it and its affect on waters suction lift?

POSSIBLE RESPONSE:

ATMOSPHERIC PRESSURE = 14.7 PSI at sea level and decreases about 1/2 PSI (.5) for each rise of 1000 feet above sea level. This affects lift about one foot per 1000 feet above sea level.

QUESTION:

How many gallons does an **IMPERIAL GALLON** equal?

POSSIBLE RESPONSE:

One **IMPERIAL GALLON** used in Canada = 1.2 U.S. gallons.

QUESTION:

At what temperature does salt water **BOIL**?

POSSIBLE RESPONSE:

Salt water **BOILS** at **226 DEGREES F**.

QUESTION:

What is the most important aspect of picking up **FROZEN HOSE**?

POSSIBLE RESPONSE:

The most important aspect of picking up **FROZEN HOSE** after a fire is **SPEED**.

QUESTION:

What is the **ELEVATION GAIN** in a 10% grade at a distance of 100 feet?

POSSIBLE RESPONSE:

A 10% grade is an **ELEVATION** of **10 FEET** in a 100 foot distance.

QUESTION:

What is the **ELEVATION LOSS** of a 15% downgrade at a distance of 100 feet?

POSSIBLE RESPONSE:

A 15% downgrade is a **DECLINE** of **15 FEET** in a distance of 100 feet.

QUESTION:

At what temperatures are the degrees of fahrenheit and centigrade **IDENTICAL**?

POSSIBLE RESPONSE:

-40 DEGREES is the only temperature that fahrenheit and centigrade are **IDENTICAL**.

QUESTION:

THEORETICALLY, 1 gallon of water at 90% cooling capacity will cool how much fire within 30 seconds?

POSSIBLE RESPONSE:

THEORETICALLY, 1 gallon of water at 90% cooling capacity will cool **100 CUBIC FEET** of fire within 30 seconds.

QUESTION:

How far will **GASES** expand?

POSSIBLE RESPONSE:

GASES will expand **INDEFINITELY!**

QUESTION:

What type of **FIRE EXTINGUISHERS** should be carried on pumpers?

POSSIBLE RESPONSE:

FIRE EXTINGUISHERS on pumper shall be suitable for class A, B and C fires, minimum size shall be:

1. Dry chemical extinguishers shall be 20 BC.
2. 2A water extinguishers shall be 2 1/2 gallon.
3. Carbon dioxide extinguishers shall be 10 BC.

QUESTION:

What are the **FIRE EXTINGUISHER** ratings and color codes?

POSSIBLE RESPONSE:

FIRE EXTINGUISHER ratings and color codes:

1. Class A fires = ordinary combustibles; green triangle.
2. Class B fires = flammable liquids; red square.
3. Class C fires = electrical; blue circle.
4. Class D fires = metal fires; yellow star.

QUESTION:

When standing near fire apparatus, in relation to the **CHARGED** hose line, where is the safest place to stand?

POSSIBLE RESPONSE:

It is safer to stand **INSIDE** of the bend in **CHARGED** hose.

QUESTION:

What is the standard friction loss **FORMULA**?

POSSIBLE RESPONSE:

$$FL = 2Q^2 + Q$$

QUESTION:

How is the **CAPACITY** of a public water system determined?

POSSIBLE RESPONSE:

The **CAPACITY** of a public water system is determined by the total amount of water it must furnish divided by the water required for domestic and industrial use in addition to the water required for the fire service.

QUESTION:

When is a water system considered **RELIABLE**?

POSSIBLE RESPONSE:

A water system is considered **RELIABLE** when it can supply the required fire flow for the number of required duration hours, with the domestic daily rate under certain emergency or unusual conditions at the maximum daily rate.

QUESTION:

What are the different types of **WATER CONSUMPTION** associated with public water systems?

POSSIBLE RESPONSE:

The types of **WATER CONSUMPTION** are:

1. Average daily consumption.
2. Maximum daily consumption.
3. Peak hourly consumption.

QUESTION:

As far as water systems are concerned, what does the term **LAMINER** refer to?

POSSIBLE RESPONSE:

LAMINER refers to turbulence.

QUESTION:

What is the minimum **DIAMETER** for a water main in residential areas?

POSSIBLE RESPONSE:

The minimum **DIAMETER** is **6 INCHES**.

QUESTION:

In comparison to a 6 inch main, what is the **CAPACITY** of an 8 inch main?

POSSIBLE RESPONSE:

An 8 inch main has **TWICE THE CAPACITY** of a 6 inch main.

QUESTION:

Name some **COMBUSTIBLE METALS**.

POSSIBLE RESPONSE:

Some **COMBUSTIBLE METALS** are:

1. Potassium.
2. Calcium.
3. Lithium.
4. Hafnium.
5. Magnesium.
6. Titanium.
7. Sodium.
8. Zinc.
9. Zirconium.

QUESTION:

What does **BLEVE** stand for?

POSSIBLE RESPONSE:

BLEVE ; **B**oiling **L**iquid **E**xpanding **V**apor **E**xplosion.

QUESTION:

What is the best method of extinguishment of **GAS FIRES**?

POSSIBLE RESPONSE:

The best method of extinguishment of **GAS FIRES** is to stop the flow of the gas.

QUESTION:

What is the **PRIMARY FUNCTION** of an engine company?

POSSIBLE RESPONSE:

The **PRIMARY FUNCTION** of an engine company, is to obtain and deliver water.

QUESTION:

What is the **SEQUENCE** of fire fighting?

POSSIBLE RESPONSE:

The **SEQUENCE** of fire fighting is:

1. Locate the fire.
2. Confine the fire.
3. Extinguish the fire.

QUESTION:

What is the number one **PRIORITY** of the fire department?

POSSIBLE RESPONSE:

The number one **PRIORITY** of the fire department is the protection of life.

QUESTION:

What are the two **BURNING MODES** of fire?

POSSIBLE RESPONSE:

FLAMING MODE : tetrahedron, 2nd phase = free burn.
SMOLDERING MODE : fire triangle, 3rd phase = smoldering.

QUESTION:

What is atmospheric **AIR** comprised of?

POSSIBLE RESPONSE:

Atmospheric **AIR** contains:

1. 21% Oxygen
2. 78% Nitrogen.
3. 1% miscellaneous gases.

QUESTION:

What are the **THREE STAGES** of a fire?

POSSIBLE RESPONSE:

The **THREE STAGES** of a fire are:

1. 1st stage = smoldering or incipient phase: fire at 1000 degrees F, room at 100 degrees F, Oxygen level at 21%.
2. 2nd stage = flame producing phase: fire and room at 1300 degrees F, Oxygen level at 15% to 21%.
3. 3rd stage = smoldering phase: fire and room at 1000 degrees F, Oxygen level at below 15%.

QUESTION:

What is the procedure for locating fire in **WALLS**?

POSSIBLE RESPONSE:

To locate fire in **WALLS: LOOK - LISTEN - FEEL**

QUESTION:

What are the different ways that heat is **TRANSMITTED**?

POSSIBLE RESPONSE:

TRANSMISSION OF HEAT:

1. Conduction = direct heat contact.
2. Radiation = in all directions where matter does not exist.
3. Convection = by air currents, usually in an upwards direction.
4. Direct flame contact.

QUESTION:

What are the **PRODUCTS OF COMBUSTION?**

POSSIBLE RESPONSE:

PRODUCTS OF COMBUSTION:

1. Fire gases. (Oxygen, Hydrogen, and Carbon)
2. Flame.
3. Heat.
4. Smoke.

QUESTION:

What is the most useful information that you can have concerning the **HAZARD** of a liquid?

POSSIBLE RESPONSE:

The most useful information that you can have concerning the **HAZARD** of a liquid is the **FLASH POINT** of the liquid.

QUESTION:

What is the order of **SPRINKLER RATINGS**?

POSSIBLE RESPONSE:

SPRINKLER RATINGS:

1. Ordinary - no color = 135 - 170 degrees F.
2. Intermediate - white = 175 - 225 degrees F.
3. High - blue = 250 - 300 degrees F.
4. Extra high - red = 325 - 375 degrees F.
5. Very extra high - green = 400 - 475 degrees F.
6. Ultra high - orange = 500 - 575 degrees F.

QUESTION:

What is the sprinkler **COVERAGE** from a sprinkler with an 1/2" orifice?

POSSIBLE RESPONSE:

Sprinkler **COVERAGE** according to occupancy hazard, with an 1/2" orifice:

1. Light = 130 to 168 square feet.
2. Ordinary = 130 square feet.
3. Extra high = 90 square feet.

QUESTION:

What should the sprinkler discharge **PATTERN** be from a sprinkler flowing 15 GPM?

POSSIBLE RESPONSE:

The sprinkler **PATTERN** should be equal to a 16 foot circle at a point 4 feet below the sprinkler head. at 15 GPM.

QUESTION:

What is the minimum **CLEARANCE** for storage from a sprinkler head?

POSSIBLE RESPONSE:

The minimum **CLEARANCE** for a sprinkler head from storage is **18 INCHES**.

QUESTION:

What is considered as the **SAFE ZONE** for wires down?

POSSIBLE RESPONSE:

The **SAFE ZONE** for wires down is one span each way.

QUESTION:

What is the danger for a firefighter directing a hose stream on wires of less than **600 VOLTS**?

POSSIBLE RESPONSE:

There is little danger to firefighters directing hose streams on wires of less than **600 VOLTS** to ground from any distance likely to be met under conditions of ordinary fire fighting.

QUESTION:

What is **OHMS LAW**?

POSSIBLE RESPONSE:

OHMS LAW is the principle that relates voltage, current and resistance:

In an electrical circuit the current varies directly with the electromotive force and inversely with resistance.

1. Amperes = volts divided by ohms.
2. Volts = amperes multiplied by ohms.
3. Ohms = volts divided by amperes.

QUESTION:

What is the maximum continues **CURRENT** that an individual may safely be subjected to?

POSSIBLE RESPONSE:

The maximum continuous **CURRENT** that an individual may safely be subjected to is **FIVE (5) MILLIAMPERES**.

QUESTION:

In relation to fire department terms, explain: **VOLTAGE, AMPERES,** and **OHMS**.

POSSIBLE RESPONSE:

In fire department terms:
1. Electrical **VOLTAGE** is like PSI. (pressure)
2. Electrical **AMPERES** is like GPM. (rate of flow)
3. Electrical **OHMS** is like FL. (resistance).

QUESTION:

What is the most hazardous **COMPONENT** of a high voltage electrical circuit?

POSSIBLE RESPONSE:

ELECTRICAL AMPERES is the most hazardous **COMPONENT** of a high voltage electrical circuit.

QUESTION:

What does the term **SUBLIME** denote?

POSSIBLE RESPONSE:

SUBLIME is when a solid changes to a vapor without passing through the liquid phase.

QUESTION:

How many **BTU'S** will one gallon of water absorb?

POSSIBLE RESPONSE:

One gallon of water will absorb **8000 BTU'S**.

QUESTION:

Explain the **ISO GRADING SCHEDULE**.

POSSIBLE RESPONSE:

ISO GRADING SCHEDULE:

1. Water supply = 40 points.
2. Fire department = 50 points.
3. Communication = 10 points.

QUESTION:

What is the maximum **DEFICIENCY** points allowed for the ISO grading schedule?

POSSIBLE RESPONSE:

The maximum **DEFICIENCY** points allowed for the ISO grading schedule is **100 POINTS**.

SECTION 8

A LIST OF QUESTIONS

A LIST OF ORAL INTERVIEW QUESTIONS:

The following list is a compilation of interview type questions that each candidate should read and be aware of. The greater number of questions that a candidate becomes familiar with, prior to the interview, the less chance for the candidate to be caught off balance during the interview.

Read the following list of questions, and formulate your own appropriate answers:

For what **REASON** do you want the position of **FIRE CAPTAIN**?

What, in your opinion, are the **DUTIES** of the **FIRE CAPTAIN**?

How have you **PREPARED** yourself for the position of **FIRE CAPTAIN**?

What makes you think that you are **MORE CAPABLE** of handling the position of **FIRE CAPTAIN**, than any of your competitors?

What part of your formal **EDUCATION**, obtained before you joined the department, do you think would help you the most as a **FIRE CAPTAIN**?

Which one or more of the **COURSES** you have taken recently do you think would help you the most as a **FIRE CAPTAIN**?

If there was no difference in the **PAY**, would you still desire to be promoted to the position of **FIRE CAPTAIN**?

What do **YOU** have to **OFFER** the department if you are promoted to the position of **FIRE CAPTAIN**?

What combination of **EXPERIENCE** and or **TRAINING** do you consider to be the best possible, for a candidate wishing to promote to the position of **FIRE CAPTAIN**?

If you were **PROMOTED** to the position of **FIRE CAPTAIN** tomorrow, how long would it be before you were ready to assume the responsibilities of the job? What, if anything, would prevent you from assuming all the responsibilities of a **FIRE CAPTAIN** immediately? Can you do anything prior to appointment to reduce this break in time?

What is a good **TRAINING** program?

In your own words, tell us what a **FIRE CAPTAIN** has to **THINK ABOUT** on the way to a fire, and what he has to do when he returns from the fire?

Is there anything at the **FIRE CAPTAIN'S** level that you have noticed is **WRONG** or **INEFFICIENT**? Could this be corrected by training? How?

In this complex structure of the fire department, where do you think the **ROLE** of the **FIRE CAPTAIN** fits?

What do you consider the most **DIFFICULT** phase of the **FIRE CAPTAIN'S** position? How have you prepared yourself for this particular phase of the **FIRE CAPTAIN'S** position?

Suppose that you knew another **FIRE CAPTAIN** was **VIOLATING** department rules, what would you do about it?

Would you say that your **SOCIAL** life would be comparatively **UNCHANGED** if you are promoted to the position of **FIRE CAPTAIN**?

What should a member of the fire department do if he discovers that the fire chief is deliberately taking home department supplies for **PERSONAL** use?

What is the greatest **MISTAKE** that you have ever made as a firefighter?

What **IMPROVEMENTS** would you suggest for the improvement of your fire department?

What is your ultimate **GOAL** with the fire department?

What do you see in the **FUTURE** of your department within the next decade?

In your opinion, what is **LEADERSHIP**?

Suppose you **DISAGREE** with a policy your acting chief has, how would you handle it?

Assume the fire chief has taken seriously ill and retires suddenly, how much personal **LOYALTY** do you owe the acting fire chief, before he proves himself?

If it is up to you to make a **DECISION** on a situation, would you decide in favor of management or in favor of personnel?

What would you do if you knew that some of the men you work with **DISLIKED** you?

Suppose that you particularly wish to win the **ESTEEM** of the men that you work with, how would you go about getting it?

Do you think it is wise to present more than one **SOLUTION** to your superior when asked for your recommendations on a certain matter?

Assume that your superior **BAWLS** you out for an error caused by a **FIRE CAPTAIN** on another shift, how would you tell him that you were not to blame?

What part of your work **EXPERIENCE**, in your opinion, fits in the most for the position of **FIRE CAPTAIN**?

In what way do you think a member of the department has a **RESPONSIBILITY** to be active in community affairs on his own time?

Do you think it is good for a member of the department to have an active **HOBBY** or two, which are completely divorced from the department?

Should **GROUPS** have more **INFLUENCE** than **INDIVIDUALS** in decisions which the department makes?

In purely **SOCIAL** conversation, **OFF-DUTY**, what **RESPONSIBILITY** does a member of the fire department have in **DEFENDING** other fire departments that are being **CRITICIZED**?

Why is it fair for government agencies to require a **LOYALTY OATH**, when industry as a rule does not?

In what ways can a member show his **LOYALTY** to the department and the local government?

Assume you are appointed to the position of **FIRE CAPTAIN** and assigned to a station where you don't know any of the members, and one of them is next below you on the list, the other men admire and respect him and believe he should have been appointed out of order. What **PROBLEMS** does this create for you and what would you do about it?

What do you think of **CRITICISM** of shifts work, other than your own?

Suppose you are asked by the **FIRE CHIEF** to address the Jr. Chamber of Commerce at 9:00 am the following day and the Jr. Chamber of Commerce is set against a department budget request for three million dollars for expansion purposes. It is up to **YOU** to change their attitude. What points would you stress in your speech?

What is your opinion of **"C" SHIFTING**?

What do you think of your departments present **EVALUATION** form?

What fire service **LITERATURE** or **BOOKS** do you have or read?

Who are the authors of the **HANDBOOK FOR FIRE PREVENTION**?

What are the **DUTIES** of the **NATIONAL BOARD OF FIRE UNDERWRITERS**?

What are the **DUTIES** of the **STATE FIRE MARSHAL** set forth?

Who is the **STATE FIRE MARSHAL**?

What is your opinion of **MUTUAL AID** and **AUTOMATIC AID**?

What do you think of open **CRITICISM** of the department

Why do **RULES** and **REGULATIONS** have to be part of the fire service?

Enumerate ten rules of the department which must be **STRICTLY ADHERED** to by all members of the department?

What is meant by **SIZE-UP**?

What is meant by **FIRE-TACTICS**?

Where would you **DIRECT** a stream, if you were combatting an **INTERIOR** fire by the **INDIRECT** method?

How would you **VENTILATE** a five story building, heavily charged with smoke from fire in the basement where paints, oil, and varnish are stored?

Describe the quickest and most effective way to get a fire company to work on a **STANDPIPE** for a fire on the 20th floor of a 21 story building?

Assume that you are the **FIRE CAPTAIN**, and upon arrival at a fire in a multi-storied building, you direct your **FIRE ENGINEER** to hook into the **STANDPIPE**. Your engineer reports that the manifold is **DAMAGED** and its impossible to connect lines to it, what would you do?

Suppose that you are the **FIRE CAPTAIN** in charge of a truck company when all other companies are on another alarm. You receive an alarm of a three story apartment house on fire and when you arrive you find the second floor on fire and exposures on both sides. **WHAT ACTIONS** are you going to take?

If while you were responding on a fire apparatus on the way to a fire and a man was knocked off and dangerously **INJURED**, what would you do?
What actions are you going to take as **FIRE CAPTAIN** in charge of a shift and you are informed by one of the men that the engineer on duty has a bottle of **LIQUOR** in his locker?

On your way to a fire you notice a **SECOND FIRE**, what would you do?

How many men are **PERMITTED** on a 15 foot, 20 foot, 30 foot, and 35 foot ladder?

You have attended many large fires, at any time did you **OBSERVE** anything that could have been **IMPROVED**?

What do you consider the greatest **WEAKNESS** in your departments **RADIO COMMUNICATIONS PROCEDURE**?

What is the **WORST FIRE HAZARD** in your City?

Tell what you know about the **WATER SERVICE** in the City?

What have you done in the area of **EDUCATION** and **TRAINING** to prepare yourself for the position of **FIRE CAPTAIN**?

What could your **EXPERIENCE** and **EDUCATIONAL** background contribute to this department in the position of **FIRE CAPTAIN**?

What do you see as the major **PROBLEM** at the **FIRE CAPTAIN'S** level on the department? How do you think it could be **IMPROVED**?

In your opinion, what is the **PRIMARY** knowledge, skills, or abilities, responsible for making a good **FIRE CAPTAIN**?

Who is **JOHN B. FREEMAN**?

What is the most important facet of the position of **FIRE CAPTAIN**?

What degree should the tractor-trailer be **JACKKNIFED**, when setting up a tractor aerial, in relation to the fire building?

How can you tell the **CAPACITY** of a pumper?

When and why is it necessary to use a **HARD SUCTION**?

What are the main causes of **DEFECTIVE** streams?

What are the five fundamental rules that govern **FRICTION LOSS** in hose?

You have just received instructions to take the apparatus that you are assigned to an conduct the **ANNUAL SERVICE TEST**, how are you going to conduct this test?

What are the five main causes of **CAVITATION** in pumps, while drafting or pumping from a hydrant?

What is the rule of thumb for the correct **SIZE OF TIP** on different length hose lays?

What is a **CHANGE OVER VALVE**, and how do you use it?

What is a **RELIEF VALVE**, and how does it work?

What is a **CHURN VALVE**, and how does it work?

What area of the **WATER SYSTEM** in your city is the least adequate and why?

Can you explain the **DIFFERENCE** between an **AUXILIARY COOLER (INDIRECT COOLER)** and a **DIRECT COOLER (RADIATOR FILL CONTROL)**?

What are the six basic principles that apply to **FLUID PRESSURE**?

Can you explain where a **BOURDON** tube is used and how it works?

Can you explain the theory of a **CENTRIFUGAL PUMP**?
According your departments practices, when do you **CHANGE OIL** in the fire apparatus?

If you pulled up to a multi-story apartment house, how would you know if it had a **DRY STANDPIPE SYSTEM**?

What is the make of **FIRE APPARATUS** that would best fit your departments needs?

Can you explain the theory of a **POSITIVE DISPLACEMENT PUMP**?

How much experience to you have as an **ACTING FIRE CAPTAIN** and have you ever been the first in **FIRE CAPTAIN** at a major fire?

Define your **IMMEDIATE SUPERIOR** in one word.

If a **FIRE HYDRANT** has limited capacity, what is advisable?

Why do you feel that **YOU** would make a **GOOD FIRE CAPTAIN**?

SECTION 9

ASSESSMENT CENTERS

FIRE SERVICE ASSESSMENT CENTERS

Fire Service **ASSESSMENT CENTERS** are a thorough and extensive technique for enhancing a Fire Departments precision in measuring existing and/or possible aptitude of a candidate for promotion.

Fire Service **ASSESSMENT CENTERS** are directed in an authentic type of presentation so that the skills for the position will manifest to the proctors by the candidates behavior.

Fire Service **ASSESSMENT CENTERS** normally take one full eight hour day to complete. (sometimes two or three eight hour days)

Assessment of promotional candidates are usually preceded by a training session for the assessors.

Assessors are selected by your particular Fire Department and the personnel office.

Assessors normally complete an intensive training course where they participate in the exercise that will be used.
Assessors will observe you in the exercise and systematically rate your performance.

Assessors normally are instructed to base their ratings and reports on what takes place within the assessment center.

During the **ASSESSMENT CENTER** don't try to be someone other than **YOURSELF**! Relax and get into the exercises by putting yourself into each situation and reacting as you normally would, not as you conceive the assessors would want you to.

Assessors will be watching your performance in terms of many various skills:

1. Oral communication.
2. Written communication.
3. Problem analysis.
4. Planning and organization.
5. Independence.
6. Interpersonal relations.
7. Organizational sensitivity.
8. Development of subordinates.
9. Persuasiveness.
10. Judgement.
11. Oral presentation.

The basic requirements for any management position will include skills in:

1. Planning.
2. Organizing.
3. Coordinating.
4. Leadership.
5. Budgeting.
6. Public relations.
7. Employee relations.
8. Personnel development.
9. Management.
10. Personal development.

During the **ASSESSMENT CENTER** you and the other candidates will take part in a series of individual and group activities. Every candidate will participate in each exercise selected for your particular **ASSESSMENT CENTER**, but not in the same order.

The **ASSESSMENT CENTER** schedule is set up so that each member of the **ASSESSMENT CENTER** team will have a chance to observe and/or interview you in one or more exercises.

The basic parts of an **ASSESSMENT CENTER** include:

1. The Oral Interview.
2. Visual Self History Presentation.
3. Oral Presentation.
4. In-Basket Exercise.
5. Group Exercise.
6. Original Writing Exercise.
7. Rating Of Counterparts.
8. Critique.

ORAL INTERVIEW

ASSESSMENT CENTERS will also have a portion set aside for an Oral Interview. Some examples as to how the are handled follows.

EXAMPLE #1:

The assessors may conduct an Oral interview as the first portion of the **ASSESSMENT CENTER**. The interview normally is structured as a typical Promotional Oral Interview. You can prepare for this portion by studying Chapters 1 through 8 of this book.

EXAMPLE #2:

Some **ASSESSMENT CENTERS**, the assessors will withdraw each candidate, one at a time, during the writing portion of the **ASSESSMENT CENTER** and conduct a short version of an Oral Interview.

EXAMPLE #3:

During some **ASSESSMENT CENTERS**, an assessor will interview each candidate sometime after the In-Basket exercise, individually. First by asking questions to find out about their background, experience, general knowledge, job knowledge, and personal knowledge, ETC.

The assessor will then interview each candidate as to the reasons why he handled particular items of the In-Basket exercise in the manner that he did.

This type of an Oral Interview can last anywhere from 1 1/2 hours to 2 1/2 hours. Normally the time is fixed before the interviews.

Chapters 1 through 8 of this book will help you for this type of interview.

VISUAL SELF HISTORY PRESENTATION

EXAMPLE:

All candidates will normally receive a 3 foot by four foot piece of paper and a felt tip ink pen.

You will be instructed to develop a graphic presentation of your personal and professional past.

You will normally get about thirty minutes to complete this graphic and prepare to present verbally it in front of the other candidates and assessors.

During this portion of the **ASSESSMENT CENTER** it is extremely revealing as to what each candidate feels is important.

The assessors will have an opportunity to observe how you and the other candidates organize your time with your work and at the same time observe how you expresses yourself verbally while standing in front of others.

ORAL PRESENTATION

This portion of the **ASSESSMENT CENTER** will give the assessors an opportunity to evaluate each candidates:

1. Verbal skills.
2. Versatility.
3. Technical knowledge.
4. Work record.
5. Educational achievements.
6. Ability to organize written material.
7. Ability to speak in front of a group.

EXAMPLE #1:

Candidates will be asked to prepare a speech of three minutes (sometimes four or five) that they will give to the group and the assessors. The group is supposed to be envisioned as the City Council, and you the candidate are playing the role of the Fire Chief.

EXAMPLE #2:

The candidates will be asked to prepare a speech to present to a local service group. (usually a five minute speech)

EXAMPLE #3:

See VISUAL SELF HISTORY PRESENTATION, presented previously.

IN-BASKET EXERCISE

During the In Basket exercise you will be asked to deal with reports, mail, notices, or any other type of routine paper work, fabricated, but applicable to the position being tested for, which supposedly has accumulated in the "In Basket" of your office.

EXAMPLES:

Candidates will receive a large envelope with blank paper and paper clips enclosed within it.

You will have from two to three hours to take the appropriate action for predetermined objectives with fifteen to thirty five items in such areas as:

1. Adjusting work schedules.
2. Management and supervisory decision.
3. Human relations.
4. Public relations.
5. Discipline.
6. **ETC.**

At the conclusion of the exercise you are to return each item to the envelope and give it to the appropriate assessor.

The assessor will use the envelope to prepare for your interview, at the completion of the exercise each candidate will be interviewed by this assessor.

The assessor will not be so much concerned with what you have accomplished with each particular item, the assessor will want to know why you choose the action that you did.

GROUP EXERCISE

There are many group exercises that are used in **ASSESSMENT CENTERS**. Some of the exercises are a group effort and others are of a competitive nature.

In many previous **ASSESSMENT CENTER** you and the other candidates will be divided into leaderless groups (usually four to six per group) to play management games. Normally for about one hour.

EXAMPLE #1:

Each group will be given a dollar amount to purchase an undetermined item for the Fire Department.

Each candidate then must decide what they think is the most important item/items within the dollar amount that they would purchase. Then try to convince the others why your selection is the most important. All the time trying to cooperate and reach an agreement.

The exercise may be stopped and each candidate may be asked to write a report evaluating his own performance and that of the other candidates.

Assessors will observe and evaluate each candidate in this exercise for:

1. Leadership.
2. Organizational ability.
3. Financial insight.
4. Thinking agility.
5. Ability under stress.

EXAMPLE #2:

Candidates will be asked to resolve a disciplinary situation involving more than one subordinate.

The groups must decide how to budget their time between each case along with resolving each particular case.

This exercise will allow the assessors to evaluate each candidates:

1. Appreciation of personnel problems.
2. Sensitivity to subordinates.
3. Handling of circumstances.
4. Reactions to others.
5. Performance within a group.

OTHER EXAMPLES:

Other situations that the group exercises could include are:

1. Command decisions.
2. Operating procedures.
3. Rules and regulations.
4. Policy and procedures.
5. Fire prevention.
6. Manpower.
7. Public relations.
8. Personnel relations.
9. Sick leave policies and problems.
10. **ETC.**

ORIGINAL WRITING EXERCISE

This portion will show the assessors each candidates ability to write.

EXAMPLE #1:

Each candidate will be given the assignment to write a speech (with a prescribed time) on any subject of their choice.

EXAMPLE #2:

Each candidate will be requested to write an review in response to a study that is critical about their Fire Chief.

RATINGS OF COUNTERPARTS

During this exercise each candidate is simply asked to rate each candidate, except for themselves.

EXAMPLE #1:

Each candidate will be given a rating sheet and asked to use the guidelines within this sheet while rating the candidates.

EXAMPLE #2:

Each candidate will be asked to rank each candidate from the most capable to the least capable.

EXAMPLE #3

Each candidate will be asked to write a character summary of each candidate.

CRITIQUE

It is recommend that each candidate contact the assessors in person or by telephone and critique his performance.

SECTION 10

SIMULATOR EXAMS

WHAT IS A SIMULATOR EXAM?

SIMULATOR EXAMS are a technique of accessing a candidates fire command practices and principles within a classroom simulator environment with the use of mechanical and audio-visual equipment producing an emergency incident. The technique evaluates the candidates size-up and fire ground tactics and strategy within a controlled environment. The test allows for various fire situations to be created and modified in response to the actions taken by the candidate along with role players. (role players explained later)

WHY HAVE SIMULATOR EXAMS?

A simulator can create various situations for the candidate that could actually take place during a real incident.

Fire situations can be modified and predicted in response to the actions taken by the candidate and role players

The simulator exam can evaluate the candidates:

1. Command skills.
2. Control and management skills.
3. Knowledge of structural conditions.
4. Knowledge of fire behavior factors as they relate to the spread and growth of fire.
5. Knowledge of target hazards.
6. Fire ground procedures.
7. Decision making methods.
8. Communication skills.
9. Radio procedures.

SIMULATOR TECHNIQUES:

Simulators incorporate the use of overhead projectors and 35 mm projectors for the reproduction of fire and or smoke. Also an amplification system for simulating radio frequencies.

The basis of the operation is the use of the 35 mm slide projected on the primary screen.

The slide may be a picture of:

1. Single family residence.
2. Apartment structure.
3. Industrial complex.
4. Commercial structure.
5. Target hazard: example lumber yard.
6. Life hazard: example hospital.
7. Water emergency.
8. Vehicle incident: autos, tankers, etc.
9. Aircraft incident.
10. Railroad incident.
11. BELVE.
12. Tank farm.
13. Virtually any type of incident!

The simulation exam will start with only the projected image of the incident location on the screen.

Additional images needed for the incident will be projected over the problem with the use of the overhead projector.

The overhead projector will show red for fire and a light grey for smoke. These images are blocked out by a light shield (usually sand) until they are needed for the problem or fire behavior. The exam proctor will create a problem by allowing the different colors to come through the light shield. He may create an assortment of visual effects in various locations.

The exam proctor can control the fire behavior to respond to the actions taken by the candidate or as instructed by the exam board. Fires may extend vertically, horizontally, create exposure fires, etc.

The candidate can restrict the fire or smoke spread by the use of his divisions of role players, thus giving the image of fire control.

The candidate and the role players utilize the communication system to simulate communication:

1. Officers.
2. Outside agencies.
3. Dispatch center.
4. ETC.

Some simulator exams may have a separate amplification system to be utilized for background noises of fire, fire engines, sirens, explosions, ETC.

Display boards, sometimes three dimensional if a projected system is used, will display where:

1. Equipment is located.
2. Hose lines are laid.
3. Hose lines are in use.
4. Ventilation takes place.
5. Entry is made.
6. ETC.

ROLE PLAYING:

Each candidate will be assigned **ROLE** for the exam. His role usually will be as **IC** (Incident Commander), but may also be:

1. Company Officer.
2. Chief Officer.
3. Dispatcher.
4. Facilitator.
5. ETC.

After the candidate is assigned his role he will be assigned role players for the other needed positions. Each of the role players will convey the responsibilities and functions of their assigned roles. They will react using sanctioned principles of fire ground operations and tactics. The candidate will be required to issue the proper orders and to respond and communicate as he would during an actual incident.

The candidate and the role players will be given a scenario of written and/or oral facts. The candidate and the role players should begin communications, actions and orders based upon their own assessment of the incident situation.

The candidate should use his own experience and knowledge that he has experienced and learned during his career.

The primary companies of the Fire Service are:

1. Engine Companies.
2. Truck and/or Ladder Companies.
3. Salvage Companies.
4. Rescue Companies and/or Paramedic Units.

The basic Firefighting unit of the Fire Service is the **ENGINE COMPANY** which usually is assigned the functions of:

1. Fire Extinguishment.
2. Exposure protection.

The primary functions of a **TRUCK COMPANY** are:

1. Ladder operations.
2. Ventilation procedures.
3. Overhaul procedures.
4. Forcible entry.
5. Rescue operations.
6. Utility control.
7. Salvage operations. (unless department has salvage company for these operations)

The primary function of a **SALVAGE COMPANY** is:
1. Salvage.

The primary function of a **RESCUE COMPANY** is:
1. On scene emergency medical procedures.

THINGS THAT YOU CAN EXPECT THE "SIMULATOR EXAM" TO CONSIST OF:

1. The use of a representative cross section of typical occupancies.
2. Handout material.
3. Realistic time tables.
4. Results to correspond to actions taken.

5. The use of slides and diagrams of local buildings and areas when possible.
6. Opportunity to familiarize yourself with the equipment.
7. A reasonable amount of information pertaining to the problem.
8. Problem will not build too fast.
9. Simulation area will not be overcrowded with people or equipment.
10. No tricks in the problem.
11. Realistic situations.
12. Problems will not escalate beyond the control of the proctors.
13. Various types of problems.

EXAM FORMAT:

The Fire Simulator Proctors will usually create a situation where you will be the first in officer.

The first in officer is expected to:
1. Take command.
2. Assume responsibility of establishing an incident command post.
3. Directing additional incoming companies.
4. Evaluating personnel.
5. Evaluating needs for apparatus and equipment.
6. Requesting additional manpower, apparatus, and equipment. (including special equipment)
7. Requesting traffic control and/or any other special services required for the incident.

At a structure fire the Fire Simulator Proctors usually will expect you to set the **INCIDENT COMMAND POST** up in the front of the structure. This will allow you to guide units coming on scene and communicate with radio dispatch. You as the Incident Command officer must remain at the Incident Command Post, therefore you must turn your company over to an acting officer.

SOME OF THE ADVANTAGES OF THE COMMAND POST ARE:

1. A stationary position.
2. Quite place to think and make decisions.
3. A vantage point.
4. Adequate lighting.
5. Area to record.
6. Radios that are more powerful.

INITIAL REPORT:

Immcdiately upon arrival of the scene of the incident make a **SIZE-UP** of the situation and report it to dispatch. This report is important to all the responding apparatus.

The description should focus on:
1. Life threatening situations.
2. Location of fire.
3. Travel direction of the fire.
4. Exposure problems.

Report should at least consist of:
1. Your location.
2. Type of incident.
3. Severity of incident.
4. What you need.

EXAMPLE OF AN INITIAL REPORT:

"Engine 82 to dispatch, we have large amounts of fire and smoke coming from the roof of a single story, wood frame, stucco exterior industrial building, about 100 feet wide by about 150 feet deep.

Significant exposure problem on the east side of the building.

Dispatch a 2nd alarm (mutual aid, automatic aid, or just ask for specific units, ie.: three additional engine companies and an additional Aerial Ladder Truck).

We are laying a line to the front of the incident, this will be the location of the Incident Command Post.

Dispatch Police Department for traffic control.

What other units are responding on this call?"

You will be expected to know what your departments initial response will be, but you may not be aware of other units out of service on another call or whatever.

ESTIMATING MANPOWER:

Fire simulator exam proctors will almost always give you a situation that requires more manpower than you will have in your first in response.

Remember that you are the Incident Commander and that you are expected to know if additional help is needed and how much! you will be expected to justify what you ask for.

A general plan to follow is to ask for the number of companies that you estimate you will need, then add one additional company just to play it safe.

The exam proctors usually will except an excuse for asking for too many companies, but will not except excuses for coming up short.

As the candidate you should have some method of estimating the number and type of companies you will need. This will help you in estimating the manpower needed and also give you a reason to explain to the proctors as to why you handled the situation the way that you did.

Discuss with officers on your Department as to the guidelines that they use for estimating fire companies in different situations.

If additional companies are needed during the exam, it is not your responsibility to determine where they come from. Your are expected to determine what is needed and then request it.

IC OPERATIONS TO BE COMPLETED:

Following is a list of some of the more important operations that the Incident Commander should oversee and follow-up to completion:

1. Life safety and rescue operations.
2. Confinement of the fire.
3. Extinguishment of fire.
4. Exposure protection.
5. Ladder operations.
6. Overhaul.
7. Ventilation.
8. Forcible entry.
9. Salvage.
10. Control of all the utilities.

ROUTING RESPONDING COMPANIES:

Fire Simulator Proctors will be watching to see that you follow your Departments policies as far as firefighting procedures and as to where you place your responding companies.

When routing your companies, remember the operations that are to be completed and the primary functions of each company!

UNUSUAL INCIDENTS:

There are many unique circumstances that you may be put into during the exam such as:

1. Hazardous materials incidents.
2. Incidents involving unfamiliar materials.
3. Incidents or circumstances unfamiliar to you.
4. Incidents demanding unique information that you do not possess.

In handling these types of incidents you can follow some of these suggestions:

1. Do not do anything that may place personnel or the public in danger.
2. Do not do anything to worsen the incident.
3. Acquire as much information regarding the incident as possible before taking any course of action.
4. Get advice from authorities that specialize in whatever the incident involves. Remember that nobody knows everything and that it is not a discredit to seek help.
5. Also be careful not to overestimate the incident.

THINGS TO KNOW WHEN COMMANDING AN INCIDENT:

1. Where to assign your company.
2. Where to set-up the Incident Command Post.
3. Priorities of incident.
4. Number of engine companies needed for fire extinguishment and exposure protection.
5. How many truck, salvage, and rescue companies you will need.
6. What special apparatus and equipment will you need.
7. The initial responding units.
8. Where to assign the first and second in engine companies. (if needed)
9. Where you intend to place other incoming engine companies.(if needed)
10. What task to assign the first and second in truck companies. (if needed)
11. Do you have an sufficient amount of units to handle all operations.
12. Your initial report to dispatch.
13. Orders for all companies that are responding.
14. If special services are needed, ie.: Utility Companies, or Police Department for traffic control, etc.

POST EXAM:

Sometimes after the simulator portion of the exam, the exam proctors may call the candidate back into the exam area and ask him to explain his:

1. Responses and actions.
2. Decisions or theories, why they were appropriate.
3. ETC.

HOW SIMULATOR EXAMS ARE GRADED:

Some things to keep in mind when taking the simulator portion of the Fire Captain exam:

1. Remember the elements of communication.
2. Have broad-minded objectives and give orders in those terms.
3. Presume subordinate officers will assume and respond to orders given in broad terms.
4. Recognize that strategy is an overall blueprint by which the crisis is handled.
5. Recognize that tactics is the attempt that is directed toward the accomplishment of the plan.
6. Sustain a concise thought process and relinquish the particulars to the Firefighters.

7. Remember large fires are somewhat like small ones, however they need more men and equipment and take more time to extinguish.
8. Remember that large fires require more fire flow and that it must be applied at greater distances.
9. Delegate control criteria based on objectives. These objectives should have the measure of quantity, quality and time.
10. Recommend that companies stay together in the sector assigned.
11. Encourage company officers to use judgement and assume a command role if need be.
12. Remember the particular objectives that may need to be met:
    - Rescue and evacuation of inhabitants
    - Life safety of Firefighters
    - Protection of exposures
    - Confinement of the fire
    - Fire extinguishment
    - Overhaul of the fire scene
    - Ventilation
    - Salvage
13. Avoid **TUNNEL VISION** in directing operations.
14. Factors for establishing objectives:
    - Construction of structure
    - Size of fire
    - Direction of fire travel
    - Occupancy classification
    - Safety of Firefighters
    - Safety of Citizens
    - Potential of the emergency
    - Priorities of the operation
    - Weather
    - traffic
    - water supply
    - Manpower
    - Equipment available
    - Resources available
15. Remember to set up a command post.
16. Inform all subordinate officers aware of command post location.
17. Initiate a strong command structure and maintain unity of command.
18. If needed redeploy companies.
19. Recognize reflex times in handling the emergency.

FIRE SIMULATOR EXAM GRADING SHEET:

During the Simulator Exam the Proctors will have a grading sheet in front of them. This sheet will consist of a list containing the components should be used by the **IC** during an incident.
These components will usually be rated as:

1. Exceptional or outstanding.
2. Good or above average.
3. Average.
4. Below average or poor.

Sometimes the rates may use a number scale such as 1 to 10 or 1 to 15, with the highest number being outstanding and the lowest number being below average or poor. The rating sheets will almost always have aa area for additional comments.

The **GRADING SHEET** will usually consist of the following components:

1. Size-Up, Initial report, 952 report:
   Evaluation of the incident.
   Clarity of orders.
   Brief with an objective.
   Notification for additional manpower.
2. Strategy and Tactics:
   Use of manpower and equipment. (on scene and incoming units and officers)
   Location of hoselines.
   Internal and external exposures covered.
   Ventilation.
3. Control of:
   Self and the problem.
4. Were objectives based on:
   Time.
   Quality.
   Quantity.
5. Productive use of order with all incoming units: With particular emergency incident.
6. Effective fire ground organization using:
   Chain of control, chain of command.
   Span of control.
   Unity of control.
   Division of manpower.
   Staging areas.
   Operations of command post.
   Coordination of decisions. (decision loop)
7. Effective communication techniques with other personnel.
8. Use of open-minded objectives, while exercising good management techniques.

SOME EXAMPLES OF SIMULATOR EXAM INCIDENTS:

REMEMBER:

1. Control incident.(give the situation direction)
2. Get "Feed Back".
3. One assignment to each company at a time.
4. Give direction not details.
5. Use equipment and manpower as needed, save the balance until in staging needed.
6. Log events as they occur.
7. Keep command available: delegate responsibilities.
8. Use special advisors as needed, if available.
9. Be aware of communications.
10. As **IC**, coordination is a primary goal.
11. During actual simulator exam you will have drawings, pictures, slides, and an amplifying system (audio and visual effects). This will assist you in your initial report and keep you apprised of situation as to what effects you are having on the incident. Don't forget to watch and listen to these devices so that you can react to the changes that are effected by your delegated assignments.

Assume that you are the first in officer for the following incidents. You will be responding with your Departments normal response of equipment and manpower with the exception that your Departments normal Incident Commander will not be responding. (unless otherwise stated)

In these exercises direction will be determined by the use of the terms; "Front, rear, left, or right of the structure". The normal terms of north, south, east, and west bearings will be of no value in these descriptions. (in an actual simulator exam it is possible that you will use either of the terms because of the visual equipment)

KEEP IN MIND THAT IN ALL OF THE FOLLOWING EXAMPLES, THAT THEY ARE ONLY EXAMPLES AND NOT NECESSARILY THE BEST OR ONLY WAY TO MANAGE THESE INCIDENTS. THE EXAMPLES ARE PRESENTED AS A METHOD TO ENCOURAGE YOU TO LOGICALLY APPRAISE EACH INCIDENT FROM BEGINNING TO CONCLUSION. THE MORE THAT YOU PRACTICE COMMANDING AN INCIDENT, THE MORE CONFIDENT YOU WILL BECOME WITH EACH INCIDENT. GO THROUGH EACH INCIDENT EXAMPLE SEVERAL TIMES AND ADD DIFFERENT CONDITIONS EACH TIME. CREATE INCIDENTS THAT BECOME MORE COMPLICATED AND SEE HOW YOU ADJUST. MANAGE EACH INCIDENT ACCORDING TO YOUR PARTICULAR FIRE DEPARTMENTS STANDARD OPERATING PROCEDURES, RESPONSE LEVELS ETC!

EXAMPLE #1

FIRE SINGLE FAMILY RESIDENCE

DESCRIPTION OF INCIDENT:

It is 0100 hrs on a Monday morning, no wind, clouds or precipitation.

You have a single story, single family dwelling, possibly occupied with two adults and two small children.

Upon arrival there is smoke and flames coming out of the kitchen area on the right side of the dwelling.

There are no exposure problems.

Special problem = upon arrival the owner of the home states to you that all of his family is accounted for except for his wife.

First alarm response = E-81, T-81, R-81.

B/C-81 is unable to respond.

THINGS TO DO DURING THIS INCIDENT:

1. Call for additional help.
2. Set up command post and give its location.
3. Call P.D. for crowd and traffic control.
4. Initiate search and rescue.
5. Ventilate.
6. Extinguish the fire.
7. Command assignments.
8. Cover vacant Fire Station with manpower and equipment.
9. Use proper timing of manpower and equipment.
10. Check for fire extension.
11. Manage utilities.
12. Forcible entry.
13. Overhaul and salvage.

POSSIBLE INITIAL REPORT:

"Dispatch, E-81 is on the scene of a occupied one story single family residence with fire and smoke showing from the right side of the structure.
We have laid a line to the front of the structure, this will be the location of the Incident Command Post.

E-81 will conduct search and rescue.

Dispatch appropriate units to cover vacated stations.

Dispatch P.D. for traffic control."

POSSIBLE FOLLOW-UP ASSIGNMENTS:

T-81 will ventilate the structure.

R-81 will relieve E-81 in search and rescue.

E-81 will attack the fire from inside structure, after ventilation is completed.

T-81 could manage utilities after ventilation is completed.

R-81 could check for extension of fire.

Continue to communicate with all units for updated reports of information and react appropriately until the incident is concluded.

This incident could escalate or the fire could be extinguished!

EXAMPLE #2
FIRE MULTI-RESIDENCE

DESCRIPTION OF INCIDENT:

It is 0400 hrs on a Saturday morning, with a slight wind blowing from the front of the structure towards the rear of the construction site. No clouds and no precipitation.

You have a two story apartment complex in the framing stage nearly complete. There is no roofing or siding on the structure. Upon arrival 75% of the first floor is involved in fire.

Exposure problems = occupied apartment buildings on the right and left side of the construction site and a stack of lumber to the rear.

Special problem = a chain link fence around the construction site.

First alarm response = E-81, E-82, T-81, R-81. B/C-81 will be delayed about 20 minutes.

THINGS TO DO DURING THIS INCIDENT:

1. Call for additional help.
2. Protect exposures with master streams.
3. Extinguish fire.
4. Set up a command post and give its location.
5. Call Police Department for traffic control and assistance in evacuating occupants from exposures.
6. Evacuate adjacent apartment buildings.
7. Set up a staging area.
8. Allow for relief of personnel and equipment.
9. Command assignments.
10. Command sectors.
11. Use proper timing with equipment and manpower.
12. Keep apprised of situation with updated reports.
13. Cover unoccupied Fire Stations with manpower and equipment.
14. Check for flying embers.
15. Set up auxiliary lighting.
16. ETC.

POSSIBLE INITIAL REPORT:

"Dispatch, E-81 is on scene, we have a two story apartment complex, in the framing stage, involved with fire over most of the first floor. There are exposures on both sides and to the rear.

We have laid a line to the front of the structure, this will be the location of the Incident Command post, E-81 crew will protect the exposure to the rear.

Call a 2nd alarm. (mutual aid plan or just ask for whatever equipment and manpower you think you need, including coverage of vacated Fire Stations)

Dispatch Police Department for traffic control and assistance in evacuating the apartment building on the right side of the incident."

POSSIBLE FOLLOW-UP ASSIGNMENTS:

"E-82 protect exposure and cool fire on right side of incident, T-81 protect exposure and cool fire on left side of incident, R-81 evacuate occupants from apartment building on left side of incident.

Second alarm response = E-83, E-84, T-82, R-82. E-83 make fire attack from rear of incident, E-84 assist E-83 in making fire attack, T-82 check surrounding area for flying embers, R-82 report to staging in parking lot directly across the street from the command post."

Continue to communicate with all units for updated reports of information and react appropriately until incident is concluded.

This incident could escalate or the fire could be extinguished!

EXAMPLE #3
FIRE MULTI-RESIDENCE

DESCRIPTION OF INCIDENT:

It is 0300 hrs on a Tuesday morning. There is no wind, clouds, or precipitation.

You have a two story, 8 unit, apartment building with fire and smoke showing.

Upon arrival you observe fire and smoke coming out of the front window of the front bottom unit.

Exposure problem = apartments above and adjacent to the right of the involved unit, along with a single family dwelling to the left of the apartment complex.

Special problem = Occupant of the unit directly above the involved unit is elderly and cannot hear without his hearing aid. He is asleep at the time of the incident.

First alarm response = E-81, E-82, T-81, R-81. B/C-81 is out of service.

THINGS TO DO DURING THIS INCIDENT:

SIZE-UP, THEN:

1. Call for additional help.
2. Search and rescue.
3. Evacuate occupants of apartment structure.
4. Evacuate occupants of exposed dwelling.
5. Protect exposures.
6. Extinguish fire.
7. Check for extension of fire.
8. Establish an Incident Command Post and give its location.
9. Call P.D. for traffic control and evacuation of occupants.
10. Set up staging.
11. Allow for relief of personnel and equipment.
12. Ventilation and forcible entry
13. Utilization of apparatus, equipment, and manpower appropriately.
14. Delegate assignments and sectors
15. Keep apprised of the situation with updated reports of information.
16. Cover vacated Fire Stations with personnel and equipment.
17. Manage utilities and add auxiliary lighting.
18. Overhaul and salvage.

POSSIBLE INITIAL REPORT:

"Dispatch E-81 is on scene, we have a two story, eight unit apartment complex, with smoke and fire showing in the lower front unit.

We have exposures problems in the apartments above and to the right side of the fire along with a dwelling to the left of the structure.

We have laid a line to the front of the structure, this will be the location of the Incident Command Post.

E-81 crew will make an interior attack on the fire through the rear window of the involved apartment."

Dispatch a 2nd alarm.

Dispatch P.D. for traffic control and for evacuation of dwelling on the left side of the incident.

POSSIBLE FOLLOW-UP ASSIGNMENTS:

R-81 conduct search and rescue of involved apartment.

T-81 evacuate remaining units of apartment complex.

E-82 protect exposure to the left of the incident.

As T-81 becomes available assign them mechanical and natural ventilation.

As R-81 becomes available assign them the task of utility management.

Second alarm response = E-83, E-84, T-82, R-82.
(in addition to vacated stations being covered)

Direct E-83, E-84, T-82, and R-82 to staging in the street 100 feet to the right of the Incident Command Post.

Continue to communicate with all units for updated reports of information and react appropriately until the incident is concluded.

This incident could escalate or the fire could be extinguished.

EXAMPLE #4

FIRE COMMERCIAL INDUSTRIAL

DESCRIPTION OF INCIDENT:

It is 2000 hrs on a Friday night, with a heavy wind blowing from the left side of the fire location towards the right side. No clouds and no precipitation.

You have a "fast print-it" shop directly in the middle of a five complex "mini-mall". It has a front entrance/exit to the parking area and a rear entrance/exit to an alley.

Flames are visible through the window of at the "fast print-it" shop.
Exposure problem = adjacent businesses, both to the right and the left of the incident.

Special problems = The last business on the right side of the incident is an occupied restaurant.(all other businesses are unoccupied)

The five businesses in this complex have a common attic.

First alarm response = E-81, E-82, T-81, R-81. B/C-81 will not be responding.

THINGS TO DO DURING THIS INCIDENT:

1. Call for additional help.
2. Protect exposures.
3. Set up command post and give its location.
4. Check for extension of fire.
5. Ventilate.
6. Extinguish fire.
7. Call P.D. for traffic control.
8. Evacuate occupants.
9. Set up staging area.
10. Allow for relief of personnel.
11. Command assignments.
12. Command sectors.
13. Cover unoccupied Fire Stations with manpower and equipment.
14. Use proper timing with equipment and manpower.
15. Keep apprised of situation with updated reports.
16. Manage utilities.
17. Forcible entry, overhaul & salvage.

POSSIBLE INITIAL REPORT:

"Dispatch E-81 is on the scene of a mini mall with the print shop directly in the middle showing fire inside its front window.

We have exposures on both sides of the incident.

We have laid a line to the front of the incident inside the parking lot. This will be the location of the command post.

E-81 crew will make a quick fire attack from the rear of the incident.

Call a second alarm.

Dispatch P.D. for traffic control and assistance for evacuation of occupants from restaurant on right end of mall."

POSSIBLE FOLLOW-UP ASSIGNMENTS:

E-82 check for extension of fire in business directly adjacent to right side of incident.

T-81 check for extension of fire in business directly adjacent to left side of incident.
R-81 evacuate occupants from restaurant at right end of mall.

Second alarm response = E-83, E-84, T-82, R-82. (in addition to vacated stations being covered)

E-83 manage utilities.

E-84, T-82, R-82 go to staging in parking lot adjacent to the Command Post.

Continue to communicate with all units for updated reports and information and react appropriately until the incident is concluded.

This incident could escalate or the fire could be extinguished!

EXAMPLE #5

FIRE COMMERCIAL INDUSTRIAL

DESCRIPTION OF INCIDENT:

It is 0500 hrs on a Saturday morning, with a strong wind blowing towards the right of the fire structure. No clouds or precipitation.

You have a warehouse 200 feet wide in the front and back and 400 feet deep, with **FIRE AND SMOKE** showing.

Upon arrival the fire is through about 75% of the roof "the building has vented itself". 25% of intact roof is at the front of the fire structure.

Exposure problems = nursery located to the rear of the fire structure, metal plating business to the right of incident and machine shop to the left of the incident.

Special problem = warehouse contains large amounts of plastic products.

First alarm response = E-81, E-82, T-81, R-81.

B/C-81 will be delayed 10 minutes.

THINGS TO DO DURING THIS INCIDENT:

1. Call for additional help.
2. Protect exposures with master streams.
3. Set-up Command Post and give its location.
4. Call P.D. for traffic control.
5. Set-up staging area.
6. Delegate assignments.
7. Delegate sectors.
8. Use apparatus, equipment, and manpower appropriately.
9. Cover vacated Fire Stations with apparatus, equipment and manpower.
10. Extinguish fire.
11. Keep apprised of situation with updated reports and information.
12. Overhaul.
13. Salvage.
14. Chemical fumes.
15. Breathing apparatus.

POSSIBLE INITIAL REPORT:

"Dispatch, E-81 is on scene, we have a warehouse fully involved in flames and " the building has vented itself".

We have exposure problems on both sides and to the rear of the fire structure.

We have laid a line to the front of the fire structure, this will be the location of the Incident Command Post.

E-81 crew will protect the exposure to the right of the fire structure.

Call a 3rd alarm.

Dispatch P.D. for traffic control"

POSSIBLE FOLLOW-UP ASSIGNMENTS:

E-82 protect exposure to the rear of the fire structure.

T-81 protect exposure on the left side of the fire structure.

R-81 report to E-81 and assist them in protecting the exposure on the right side of the fire structure.

3rd alarm response = E-83, E-84, E-85 T-82, R-82, along with B/C-82.

E-83 make an exterior attack of the fire from the front of the structure.

T-82 make an exterior attack of the fire from the right side of the fire structure, using your ladder pipe.

B/C-81 upon your arrival take charge of sector to the rear of the fire structure.
B/C-82 upon your arrival take charge of sector to the right of the fire structure.

E-84, E-85, and R-82 report to staging located in the parking lot directly across the street from the Incident Command Post.

Continue to communicate with all units for updated reports and information and react appropriately until the incident is concluded.

EXAMPLE #6

HAZARDOUS MATERIALS INCIDENT

DESCRIPTION OF INCIDENT:

It is 1400 hrs on a Friday afternoon with a breeze blowing towards the left of the incident. There are no clouds or precipitation.

You have a vehicle tanker located in a parking lot, 200 feet to the right of a restaurant and a storm drain.

The tanker, which has a placard marked with the numbers **1203** is leaking a liquid on to the ground.

Upon arrival there is **NO FIRE**, but a substantial amount of liquid is on the ground.

Exposure problem = vehicles located in the area of the spill, a storm drain to the left of the spill, and the restaurant to the left of the spill.

Special problems = the parking lot has a slight decline towards the restaurant and storm drain.

First alarm response = E-81, E-82, T-81, R-81.

B/C-81 is unavailable.

THINGS TO DO DURING THIS INCIDENT:

1. Call for additional help and delegate assignments.
2. Evacuate area and restaurant.
3. Protect exposures.
4. Identify, stop and dike leaking liquid.
5. Call P.D. for traffic control and assistance in evacuation.
6. Call city yard for a truck load of sand. (special equipment)
7. Cover vacated Fire Stations with manpower and equipment.
8. Establish an Incident Command Post and give its location.
9. Utilization of manpower and equipment appropriately.
10. Set up staging area.
11. Keep apprised of situation with updated reports of information.
12. Notify proper authorities for chemical clean up.
13. Pick up and discard substance.
14. Prevention of ignition.
15. Foam.

POSSIBLE INITIAL REPORT:

"Dispatch E-81 is on scene, we have a tanker leaking an undetermined liquid onto the ground of the parking lot 200 feet to the right of a restaurant.

There is a possible exposure problem with the restaurant along with two vehicles in the area.

We have laid a line to the right of the tanker, in the parking lot, this will be the location of the Incident Command Post.

E-81 crew will protect the area with charged foam lines.

Call a 2nd alarm.

Dispatch P.D. for traffic and crowd control.

Dispatch a truck load of sand from the city yard."

POSSIBLE FOLLOW-UP ASSIGNMENTS:

Dispatch, the leaking liquid has been identified as **GASOLINE**.

R-81 evacuate the parking lot.

E-82 Evacuate the restaurant.

T-81 dike the leaking liquid in an area between the storm drain and the tanker.

Second alarm response = E-83, E-84, T-82, R-82.

E-83 spot hydrant in area of restaurant and standby.

E-84 control the liquid leak on the tanker.

T-82, R-82 report to staging located 100 feet to the right of the Incident Command Post, in the parking lot.

Continue to communicate with all units for updated reports and information and react appropriately until the incident is concluded.

This incident has the potential of major consequences or it could be just a clean-up problem.

SOME ADDED PROBLEMS TO CONSIDER FOR EACH EXAMPLE:

1. Delayed responses of apparatus and equipment from traffic or whatever.
2. Communication problems such as radio failure.
3. Weather changes.
4. Loss of manpower due to injuries etc.
5. Loss of apparatus and equipment due to breakdowns, etc.
6. Water supply problems.
7. Broken supply lines.
8. Explosions.
9. Problems with the public.
10. Simultaneous incidents.
11. Access problems.
12. ETC.

INDEX

INDEX

A

B

C

D

D
continued

E

F

F
continued

G

H

H

continued

I

J

K

L

L

continued

M

N

O

O

continued

P

Q

R

R

continued

S

S
continued

T

U

V

W

Z